Mathematical Models for Decision Making with Multiple Perspectives
An Introduction

Maria Isabel Gomes

Centre of Mathematics and Applications
Department of Mathematics
Nova School of Science and Technology, Caparica, Portugal

Nelson Chibeles Martins

Centre for Mathematics and Applications
Department of Mathematics
Nova School of Science and Technology, Caparica, Portugal

CRC Press
Taylor & Francis Group
Boca Raton London New York

CRC Press is an imprint of the
Taylor & Francis Group, an **informa** business

A SCIENCE PUBLISHERS BOOK

First edition published 2022
by CRC Press
6000 Broken Sound Parkway NW, Suite 300, Boca Raton, FL 33487-2742

and by CRC Press
4 Park Square, Milton Park, Abingdon, Oxon, OX14 4RN

© 2022 Taylor & Francis Group, LLC

CRC Press is an imprint of Taylor & Francis Group, LLC

Library of Congress Cataloging-in-Publication Data (applied for)

ISBN: 978-0-367-44074-9 (hbk)
ISBN: 978-1-032-16845-6 (pbk)
ISBN: 978-1-003-01515-4 (ebk)

DOI: 10.1201/9781003015154

Typeset in Times New Roman
by Innovative Processors

Preface

Our primary goal for this textbook is to have, within a single volume, decision-making mathematical methods that are usually scattered over several books. We explore topics from single to multi perspective decision making to present the reader (the student) different but related concepts starting on simpler methodologies and moving towards the more complex ones, thereby smoothing the learning curve.

We do not intent to be exhaustive in any of the addressed topics. Moreover, we have been mindful with regards to which mathematical details to include in each chapter. The purpose of this textbook is to present a course providing students with the first approach on how to make decisions in a systematic (*mathematical*)way, to alert them about the methods limitations and how to handle them. Finally, we wish to provide some guidance on where the readers can look for mathematical theoretical background in case they feel the need to. Questions regarding the psychology of decision making and the influence of emotions when making decisions are out of the scope of this course.

As a pre-requisite, readers must be knowledgeable of basic matrix algebra and model building in linear optimization, and on the graphical approach to solve two-variable linear optimization problems. Problems with a larger number of variables should be solved with an optimization software. We provide a few models programed in GAMS language. However, all the proposed problems can be easily solved by the Excel solver tool.

The book is structured into three parts: single-criterion decision making, multi-criteria decision making and multi-objective optimization.

In the first part, Chapter 1 focus on single-criterion decision making. This chapter sets the basis for the remaining course. It allows students to get used to decision making language and introduces them to some basic techniques and different problem context (uncertainty and risk). Chapter 2 addresses sequence decision making by the means of a decision tree approach. We also discuss ways to model subjectivity and how it can be incorporated when making decisions (e.g. the concept of utility).

The second part addresses methods dealing with several attributes in a discrete alternative context. Four methodologies are studied representing different modeling philosophies: compensatory (SMART and TOPSIS), non-compensatory

(ELECTRE methodology) and hierarchical (AHP philosophy). Chapter 3 presents the SMART (Simple Multi-Attribute Rating Technique) methodology. Being an easily understandable method, we selected it as the first one to be presented to students. Special attention will be given to the value tree and attribute definition, on how to translate attributes into criteria, on alternatives dominance and when the dominance is a tool to reduce the problem dimension. Weights and the drawbacks of compensatory methods will be discussed. This chapter ends with the presentation of sensitivity analysis techniques for both two and three-criteria cases. For the latter, the TRIDENT method is presented. To the best of our knowledge, this method has not yet been a subject of study in a textbook. In chapter 4, the ELECTRE (ÉLimination Et ChoixTraduisant la RÉalité) methodology is presented. Our teaching experience says that, in opposition to SMART, students do not easily follow the ELECTRE methodology. Therefore, we prefer to present ELECTRE I as an introduction to the methodology which allows us to address the underlying ideas. As a more complete approach, the choice goes for the ELECTRE III. Concerning hierarchical methods (chapter 5), we have chosen to address AHP (Analytical Hierarchical Process) methodology since it is an approach frequently used in social sciences. Although several authors strongly criticized its theoretical foundations, we believe that addressing it in this course will make students aware of these issues, if later they prefer it for a particular problem. The second part of this textbook ends with chapter 5 dedicated to TOPSIS (Technique for Order of Preference by Similarity to Ideal Solution). Although of the same category as SMART, we decided to present it last since TOPSIS minimizes the distance to an ideal solution making, therefore, the bridge to the last part of this book, the multi-objective optimization. Moreover, given the simplicity of the calculations of TOPSIS, we present the Fuzzy TOPSIS model as an introduction on decision making in fuzzy environments.

The third part addresses the multi-attribute decision making with alternatives available in a continuous space. Given space and time limitation, we focus only on linear problems. The first chapter (chapter 7) presents us with details of the fundaments of multi-objective programming: what is it? what are the main differences regarding single objective optimization? These questions allow us to introduce the concepts of efficiency and Pareto optimality. A special remark is made regarding the consequences of having weakly efficient solutions. After this introductory part, chapters 8 to 10 focus on techniques to reach a compromise solution. Our selection gives students an overview of methods where the decision maker's preferences are modeled by different perspectives (*a priori*, *a posteriori* or in progressive articulation). Chapter 8 is dedicated to feasible region reduction methods such as the lexicographic approach and the ε-constraint method. Both methods are first described as *a priori* methods so that a compromise solution can be reached. Then ε-constraint method is presented as *a posterior* method (a tool to approximate the Pareto front). Given its inability to identify weakly efficient solutions, the AUGMECON approach is suggested as a tool to overcome this limitation. Chapter 9 addresses weighted sum and L_p-distance minimization models. For the former, several re-scaling techniques are discussed, while for the latter, special attention is given to L_1, L_2 and L_∞ distances since these can be modeled as linear or quadratic models and can be "easily" solvable to optimality by commercial software. Chapter 10 focuses on the iterative method STEM (STEp Method) which allows the decision maker to evaluate the compromise solution and re-evaluate her/his decisions until feeling pleased with the final solution, a progressive articulation method. Other methods are briefly described to show the diversity of approaches that have been developed over the years. Chapter 11 provides an introduction to Goal Programming with the presentation of its two classical models: the Preemptive model and the Archimedean model. Two other models are presented as examples.

This book ends with a short chapter as appendix where the most important concepts are summarized, providing support to chapters that make use of them. Also, the chapters 3 to 6 the baseline example is described.

Maria Isabel Gomes
Nelson Chibeles Martins

Contents

Single Criterion Decision Making

1.1 Introduction

Living is all about making decisions. Every time we need to do something, from the moment we wake up in the morning, we make decisions. Whether it is in our professional or in our personal life we always want to choose the most appropriate, the most adequate, or the best option from a plethora of the ones that the Universe happens to put in front of us. But how do Humans choose? How do we make decisions? Many times, even if we do not fully understand why we have made a certain decision, making a decision is something very easy. The most interesting option simply and clearly shines brightly like a beacon for our Mind's Eye. In other situations we are racked by doubt and can not clearly see what the most advantageous choice is. These difficulties can arise mainly from the two following factors: we are not able to assess the total consequences of our choices due to the uncertain nature of those choices' consequences; and sometimes one choice will provide the best outcome, from one point of view, but it will ensure a really unpleasant consequence when assessed from a different, although also very important, perspective.

The way a Human being makes decisions has been studied and explored by Psychology for decades. This science has been trying to explain and codify the human rational mechanisms behind the decision making processes.

Parallelly, Operational Research has been developing mathematical and logical techniques that support a systemization approach to decision making problems, usually, the choice of the best alternative from a set of available ones. These Techniques together form what is known as Decision Theory.

This chapter introduces Decision Theory, one Operational Research area. Section 1.2 will focus on Single Criterion Decision Making problems and will present several methodologies to solve them. Section 1.2.1 and 1.2.2 will, respectively introduce Decision Problems under Uncertainty and under Risk. The chapter ends with some final remarks and a set of exercises covering all topics.

1.2   Single Criterion Decision Making Methodologies

A Single Criterion Decision Making (SCDM) Problem involves a Decision Maker (DM), or a group of Decision Makers, facing a finite set of n strategies or alternatives (E_i) classified according to m possible events, states of Nature or attributes (θ_j). A Decision problem thus defined can be represented in a matrix, denominated by the Pay-offs Matrix, as the one presented in

Table 1.1: Pay-offs Matrix

		Events, States of Nature or Attributes			
		θ_1	θ_2	...	θ_m
	E_1	v_{11}	v_{12}	...	v_{1m}
Strategies or Alternatives	E_2	v_{21}	v_{22}	...	v_{2m}
	$\vdots$	$\vdots$	$\vdots$	$\vdots$	$\vdots$
	E_n	v_{n1}	v_{n2}	...	v_{nm}

The values v_{ij} are expected values conditioned by the random events that will happen in the future and by the adopted choice when the Decision is being made. These can be in an increasing scale, like profits, sales, satisfaction or any other beneficial measure; or in a decreasing scale, like costs, penalties, or other detrimental indicator.

All topics approached in this chapter will be framed by an example that reflects the most common characteristics and typical situations involving a SCDM problem.

Example

The creative director of the Minerva Department Store is planning to launch a marketing campaign supporting the company's brand new sportswear line. Her team proposed four different possible campaigns (I, II, III and IV). The impact of each of the proposed campaigns will depend on the National Olympic Teams' performance during the next Olympiad, an event that is happening soon. The mentioned impacts are presented in Table 1.2.

Table 1.2: Example's Pay-offs Matrix

		Team's Performances			
		Very Low	Low	Median	High
	I	100	100	100	100
Campaigns	II	38	25	43	123
	III	75	103	175	98
	IV	25	190	190	25

The director would like to choose the most adequate campaign, which means the campaign that will provide the highest impact.

Usually, in colloquial language, the terms "uncertainty" and "risk" are frequently used indifferently. However the two concepts refer to very different contexts. Frank Knight (Knight 1921) is one of the first authors distinctly defining the two notions. According to Knight, in a Risk Context it is possible to quantify the involved randomness. Meaning, it is possible to estimate the probabilities associated with each possible random outcome. In other words, the probability distribution of that random event. When that is not possible then the event will happen in a pure Uncertainty Context.

1.2.1 Decision Making in an Uncertainty Context

In a Decision Making process in the Uncertainty Context, or Choice under Uncertainty, it is not possible, plausible or desirable to know the probabilities associated with each random outcome. Sometimes that happens because the estimation of those probabilities would need information that is very expensive or difficult or even dangerous to obtain. But, on other occasions, the information is simply unavailable. Nevertheless, the Decision Maker still desires to take the most suitable option in spite of the information void. There are several approaches that provide some support to a DM in this situation:

1. Pessimistic Criterion

Also known as the Wald[1] Criterion or the Minimax Criterion. It assumes the DM is the quintessential pessimist, meaning that the DM expects to get the worst possible outcome no matter which choice is made. With this rationale the Pessimistic Criterion advises that the DM should choose the best option from the worst case scenarios.

In an increasing scale situation, the worst case for Strategy i corresponds to the minimum of the possible outcomes' values for this strategy: $min(v_{11}, v_{12}, ...,v_{1m})$ as is illustrated in Table 1.3.

And the recommended Strategy is the one that maximizes these worst case scenarios:

$$E_k: \max_i (\min_j v_{ij})) = \min_j (v_{kj})$$

That is why this criterion is also known as the Minimax Criterion. This criterion is preferred by the Risk adverse DM. Even in an Uncertainty Context they try to minimize the maximal possible loss.

---

[1] Abraham Wald was a Hungarian Mathematician with innumerous contributions in Statistics, Econometrics and Decision Theory, among others. He was responsible for the XX Century reawakening of Decision Theory, reintroducing the notions of Decision under Uncertainty and under Risk and introducing the Minimax Criterion. His works during and after World War II are considered seminal in Operational Research.

Table 1.3: Pessimistic Criterion

		Events, States of Nature or Attributes			Pessimistic
		θ_1	θ_2	...	θ_m	Criterion
	E_1	v_{11}	v_{12}	...	v_{1m}	$\max_j (v_{1j})$
Strategies or Alternatives	E_2	v_{21}	v_{22}	...	v_{2m}	$\max_j (v_{2j})$
	$\vdots$	$\vdots$	$\vdots$		$\vdots$	$\vdots$
	E_n	v_{n1}	v_{n2}	...	v_{nm}	$\max_j (v_{nj})$

Example

In the example, Campaign I has the same impact (100) independently of the Team's performance in the sport event, therefore: $\min_j (v_{1j}) = 100$. For campaign II, $\min_j (v_{2j}) = min\,(38, 25, 43, 123) = 25$. Table 1.4 presents $\min_j (v_{nj})$ for each strategy $E_i \in \{$I, II, III, IV$\}$.

Table 1.4: Example – Pessimistic Criterion

		Team's Performances				Pessimistic
		Very Low	**Low**	**Median**	**High**	Criterion
	I	**100**	**100**	**100**	**100**	**100**
	II	38	25	43	123	25
Campaigns	III	**75**	103	175	98	**75**
	IV	**25**	190	190	**25**	**25**

The maximum value from the worst case scenarios is $(100, 25, 75, 25) = 100$ that corresponds to Strategy I, so that this strategy is the one recommended by the Criterion.

It is a common mistake to assume the Pessimistic Criterion will recommend the strategy that minimizes the worst case's values but bear in mind the DM is a rational person and not particularly masochistic. When faced with a list of worst case situations the DM will naturally pick the best one.

If the conditional values were in a decreasing scale than the recommended strategy would be the one that minimizes the worst case scenarios, *i.e.*:

$$E_k : \min_i (\max_j v_{ij}) = \max_j (v_{kj})$$

2. Optimistic Criterion

Also known as the Hurwicz[2] Criterion or the Maximax Criterion, this approach will assume the DM is the quintessential Optimist. The DM expects to get the

---

[2] Leonid Hurwicz was a Polish-American Mathematician and Economist who researched Game Theory. In Decision Theory Hurwicz combined Wald's works with some ideas previously proposed in the XIX by Pierre-Simon Laplace and introduced the Optimistic DM's point of view in a Decision Making under Uncertainty process. He proposed the first version of what would later become known as the Savage Criterion. Hurwicz, and his colleagues Maskin and Myerson won the 2007 Nobel Prize in Economic Sciences.

best possible outcome no matter the chosen strategy. These decisions are made considering the best possible scenarios for each strategy and the Criterion recommends the very best of the best ones.

In an increasing scale situation the best case for Strategy i corresponds to the maximum of the possible outcomes' values for this strategy: $max\ (v_{11}, v_{12}, ...,v_{1m})$ as illustrated in Table 1.5.

Table 1.5: Optimistic Criterion

		Events, States of Nature or Attributes				Optimistic Criterion
		θ_1	θ_2	...	θ_m	
Strategies or Alternatives	E_1	v_{11}	v_{12}	...	v_{1m}	$\max_{j}(v_{1j})$
	E_2	v_{21}	v_{22}	...	v_{2m}	$\max_{j}(v_{2j})$
	$\vdots$	$\vdots$	$\vdots$		$\vdots$	$\vdots$
	E_n	v_{n1}	v_{n2}	...	v_{nm}	$\max_{j}(v_{nj})$

And the recommended Strategy is the one that maximizes these best case scenarios:

$$E_k : \max_{i} (\max_{j} v_{ij})) = \max_{j}(v_{kj})$$

And, obviously, that is why this criterion is also known as the Maximax Criterion. This criterion is preferred by the Risk taking DM. Even in an Uncertainty Context they aim to maximize their gains despite being even under the prospect of incurring great potential losses.

Example

In the example, Campaign I has the same impact (100) independently of the Team's performance, therefore: $\max_{j}(v_{1j}) = 100$. For campaign IV $\max_{j}(v_{2j}) = max\ (25, 190, 190, 25) = 190$. Table 1.6 presents $\max_{j}(v_{nj})$ for each strategy $E_i \in \{I, II, III, IV\}$.

Table 1.6: Example – Optimistic Criterion

		Team's Performances				Optimistic Criterion
		Very Low	**Low**	**Median**	**High**	
	I	100	100	100	100	100
	II	38	25	43	123	123
Campaigns	III	75	103	175	98	175
	IV	25	190	190	25	190

The maximum value from the best case scenarios is $(100, 123, 175, 190) = 190$ which corresponds to Strategy IV, such that this will be the strategy recommended by this Criterion.

If the conditional values were in a decreasing scale then the recommended strategy would be the one that minimizes the best case scenarios, *i.e.*:

$$E_k: \min_i (\min_j v_{ij}) = \min_j (v_{kj})$$

Despite their simplicity these two methods can be highly unrealistic because the majority of people, and particularly the DM, are neither totally optimistic nor totally pessimistic. The next Criterion tries to consider all possible middle ground attitudes for the DM.

3. Savage Criterion

The Savage[3] Criterion is also known as the Hurwicz Criterion (see[2] on page 4) in the literature because it was originally proposed by Hurwicz, but the most well known version is actually owed to Leonard Savage. It uses a weighted sum to combine both perspectives: the Optimistic and the Pessimistic Criteria. It recommends a Strategy according to the DM's attitude relative to optimism/pessimism.

Consider α the DM's Optimism level. This parameter can take any value in the interval, $[0, 1]$ where $\alpha = 0$ represents a total pessimistic DM while $\alpha = 1$ corresponds to a totally optimistic one. $\alpha \cong 0.5$ corresponds to a Neutral DM. The DM's attitude can be represented by the most adequate Optimism Level. A Very Pessimistic DM will correspond to α close to zero, while a Slightly Optimistic DM can be represented by α around 0.6.

S_i will be a weighted sum for Strategy i combining both the best and worst case using α, and $1 - \alpha$ as weights as presented in Table 1.7.

Table 1.7: Savage Criterion

	Team's Performances				Pessimistic Criterion	Optimistic Criterion	Savage Criterion
	θ_1	θ_2	$\cdots$	θ_n			
E_1	v_{11}	v_{12}	$\cdots$	v_1	$\min_j (v_{1j})$	$\max_j (v_{1j})$	$S_1 = \alpha \max_j (v_{1j}) + (1-\alpha)\min_j (v_{1j})$
Strategies E_2	v_{21}	v_{22}	$\cdots$	v_2	$\min_j (v_{2j})$	$\max_j (v_{2j})$	$S_2 = \alpha \max_j (v_{1j}) + (1-\alpha) \min_j (v_{2j})$
$\vdots$	$\vdots$	$\vdots$		$\vdots$	$\vdots$	$\vdots$	$\vdots$
E_n	v_{n1}	v_{n2}	$\cdots$	v_n	$\min_j (v_{nj})$	$\max_j (v_{nj})$	$S_n = \alpha \max_j (v_{nj}) + (1-\alpha) \min_j (v_{nj})$

The Savage equations represent non-decreasing lines that can be represented in a graphic. The method recommends the strategy with the highest value of $S_k(\alpha)$ for a given Optimism level α.

---

[3] Leonard Savage was an American mathematician and statistician who proposed the theory of subjective and personal probabilities still used in Bayesian Statistics and Game Theory. In Decision Theory, Savage proposed a variation and some corrections to Hurwicz Criterion (see[2] on page 4).

Note that if the conditional values v_{ij} are represented in a decreasing scale then the Savage Equations will be the negative sloped linear equations:

$$S_k = \alpha \min_j (v_{kj}) + (1-\alpha) \max_j (v_{kj})$$

And the method will recommend the strategy with the lowest $S_k(\alpha)$.

Example

Table 1.8 presents the Savage equation for each Example's alternative.

Table 1.8: Example - Savage Criterion

		Events				Pessimistic Criterion	Optimistic Criterion	Savage Criterion
		Very Low	Low	Median	High			
	I	100	100	100	100	100	100	$S_I = \alpha100 + (1-\alpha)100$
Campaigns	II	38	25	43	123	25	123	$S_{II} = \alpha123 + (1-\alpha)25$
	III	75	103	175	98	75	175	$S_{III} = \alpha175 + (1-\alpha)75$
	IV	25	190	190	25	25	190	$S_{IV} = \alpha190 + (1-\alpha)25$

The equations can be simplified:

$$S_I = \alpha100 + (1-\alpha)100 = 100$$
$$S_{II} = \alpha123 + (1-\alpha)25 = 25 + 98\alpha$$
$$S_{III} = \alpha175 + (1-\alpha)75 = 75 + 100\alpha$$
$$S_{IV} = \alpha190 + (1-\alpha)25 = 25 + 165\alpha$$

Figure 1.1 shows the Savage Equations for the Example's strategies. As can be seen above, there are three different strategies that can have the highest: $S(\alpha)$: S_I, S_{III}, and S_{IV}, depending on the value for α.

The Optimism levels where two relevant lines intersect are known as the indifference levels and correspond to the situation where it will be indifferent for the DM to choose between either strategies. Therefore, the next step is to find those intersections.

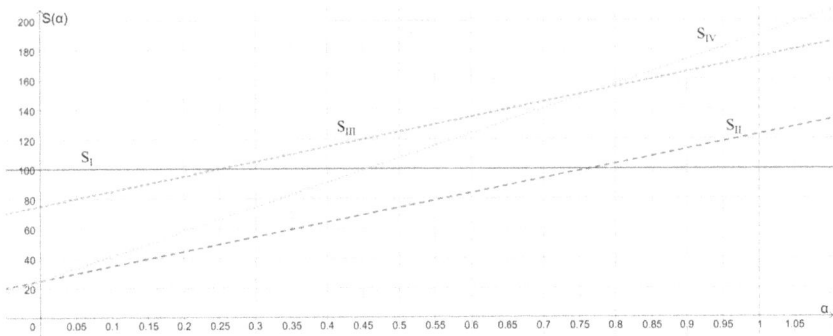

Figure 1.1: Example – Graphic with Savage Equations

Considering: $S_I = S_{III}$:

$$100 = 75 + 100\alpha \Leftrightarrow 100\alpha = 25 \Leftrightarrow \alpha = \frac{25}{100} = 0.25$$

The first indifference level is $\alpha_1 = 0.25$, therefore, strategy I will be recommended to every DM with $\alpha < 0.25$. And that corresponds to a Very Pessimistic DM.

Considering $S_{III} = S_{IV}$:

$$75 + 100\alpha = 25 + 165\alpha \Leftrightarrow 65\alpha = 50 \Leftrightarrow \alpha = \frac{50}{65} = 0.77$$

The second indifference level is $\alpha_2 = 0.77$ meaning that strategy III is recommended for DM with $0.25 < \alpha < 0.77$. And that corresponds to DMs who are moderately Pessimistic; slightly Pessimistic; Neutral; slightly Optimistic and moderately Optimistic.

Strategy IV should only be recommended to DM with $\alpha > 0.77$, which are Very Optimistic.

Notice that Strategy II will not be recommended to any kind of DM.

Although being much more realistic than both the previous Criteria, Savage Criterion nevertheless has a particular flaw that may lead to irrational recommendations. Because the Savage Equations only incorporate the best and worst cases the majority of the pay-offs' information is ignored and some of these pay-offs could have provided very relevant data.

Example

Consider the SCDM problem with a Pay-offs matrix as presented in Table 1.9, where the conditional values represent benefits.

Table 1.9: Example 2

	θ_1	θ_2	θ_3	θ_4	θ_5
A	100	1	1	1	0
B	99	98	98	98	1

The Savage Equations for both strategies are:

$$S_A = \alpha 100 + (1 - \alpha)0 = 100\alpha$$

$$S_B = 99\alpha + (1 - \alpha)1 = 98\alpha + 1$$

The indifference Optimism level will be the solution to $S_A = S_B$:

$$100\alpha = 98\alpha + 1 \Leftrightarrow 2\alpha = 1 \Leftrightarrow \alpha = 0.5$$

That means that all Optimistic DM will prefer strategy A, while all Pessimistic ones will prefer B. But should it be that simple? Strategy A has four very bad outcomes out of the possible five while strategy B has the opposite situation. Even without knowing the probabilities associated with each outcome one can not help wondering if choosing A over B makes sense to anyone.

>9

4. Equiprobable Criterion

Also called the Laplace[4] Criterion, this criterion simply assumes that all random events conditioning the outcomes of a SCDM problem have an equal probability of happening. Thus the method recommends the strategy that maximizes the Expected conditional value, if these values use an increasing scale. Symmetrically, if the scale is decreasing, then the strategy that minimizes the Expected value will be recommended. In both cases, due to the equiprobability, the Expected conditional value of a strategy corresponds to the simple average of all conditional values for that strategy.

Table 1.10: Laplace Criterion

		Events, States of Nature or Attributes			Equiprobable Criterion
		θ_1	θ_2	...	θ_m	
	E_1	v_{11}	v_{12}	...	v_{1m}	$\dfrac{\sum_{j=1}^{m} v_{1j}}{m}$
	E_2	v_{21}	v_{22}	...	v_{2m}	$\dfrac{\sum_{j=1}^{m} v_{2j}}{m}$
Strategies or Alternatives	$\vdots$	$\vdots$	$\vdots$	$\vdots$	...
	E_n	v_{n1}	v_{n2}	...	v_{nm}	$\dfrac{\sum_{j=1}^{m} v_{nj}}{m}$

Example

Table 1.11 presents the results of applying the Equiprobable Criterion to the example Pay-offs matrix.

The recommended Campaign is *III* with the highest average impact, 112.75. It is also a very simple criterion and is often used although it is not considered realistic. How often will the possible outcomes have equal probabilities?

5. Opportunity Cost (Regret) Criterion

This criterion tries to minimize the highest possible regret the DM will suffer if a non-optimal choice is taken.

The Opportunity Cost of a strategy conditioned by a possible random event (oc_{kj}) corresponds to the difference between the corresponding v_{kj} and the best possible v_{ij} for that random event, *i.e.*:

---

[4] Pierre-Simon Laplace was a French scholar who lived from the second part of the XVIII century to the first quarter of the XIX. He left us innumerous contributions in mathematics, philosophy, physics, astronomy, engineering and statistics. Bayesian Statistics has its roots in Laplace's works.

Table 1.11: Example – Laplace Criterion

		Team's Performances				Equiprobable Criterion
		Very Low	**Low**	**Median**	**High**	
	I	100	100	100	100	100
Campaigns	II	38	25	43	123	$\dfrac{38 + 25 + 43 + 123}{4} = 57.25$
	III	75	103	175	98	$\dfrac{75 + 103 + 175 + 198}{4} = 112.75$
	IV	25	190	190	25	$\dfrac{25 + 190 + 190 + 25}{4} = 107.5$

$$oc_{kj} = \max_i (v_{ij}) - v_{kj}$$

Table 1.12 presents the Opportunity Cost Matrix corresponding to a generic Payoffs Matrix when the conditional values were measured in an increasing scale.

Table 1.12: Opportunity Cost Matrix

		Events, States of Nature or Attributes			
		θ_1	θ_2	...	θ_m
	E_1	$(v_{i1}) - v_{11}$	$(v_{i2}) - v_{12}$	...	$(v_{im}) - v_{1m}$
Strategies or Alternatives	E_2	$(v_{i1}) - v_{21}$	$(v_{i2}) - v_{22}$	...	$(v_{im}) - v_{2m}$
	$\vdots$	$\vdots$	$\vdots$	$\vdots$	$\vdots$
	E_n	$(v_{i1}) - v_{n1}$	$(v_{i2}) - v_{n2}$	...	$(v_{im}) - v_{nm}$

On the other hand, if the conditional values used a decreasing scale then the Opportunity Cost of strategy E_k according to Event θ_j will be $oc_{kj} = v_{kj} - \min_i(v_{ij})$.

In both situations, though, the Opportunity Cost will be using a decreasing scale and, therefore, will have to be minimized.

Example

Table 1.13 presents the Opportunity Cost Matrix corresponding to the example's Pay-offs Matrix (check Table 1.2).

It is easier to build the Opportunity Cost Matrix column-wise because the maximal conditional values for each column are used in the calculations.

For random event θ_1 = "Very Low Performance", the best choice is campaign I with an impact of 100:

$$\max_i (v_{i1}) = \max(100, 38, 75, 25) = 100$$

Table 1.13: Example's Opportunity Cost Matrix

		Team's Performances			
		Very Low	**Low**	**Median**	**High**
Campaigns	I	0	90	90	23
	II	62	165	147	0
	III	25	87	15	25
	IV	75	0	0	98

Therefore the first column of the Opportunity Cost Matrix will be:

$$oc_{I1} = 100 - v_{I1} = 100 - 100 = 0$$
$$oc_{II1} = 100 - v_{II1} = 100 - 38 = 62$$
$$oc_{III1} = 100 - v_{III1} = 100 - 75 = 25$$
$$oc_{IV1} = 100 - v_{IV1} = 100 - 25 = 75$$

For random event θ_4 = "High Performance", the best choice is campaign I with an impact of 123:

$$\max_i (v_{i4}) = \max(100, 123, 98, 25) = 123$$

Therefore the fourth column of the Opportunity Cost Matrix will be:

$$oc_{I4} = 123 - v_{I4} = 123 - 100 = 23$$
$$oc_{II4} = 123 - v_{II4} = 123 - 123 = 0$$
$$oc_{III4} = 123 - v_{III4} = 123 - 98 = 25$$
$$oc_{IV4} = 123 - v_{IV4} = 123 - 25 = 98$$

Note that every column has a null position, corresponding to the strategy with the best possible outcome for that column/event in the Pay-offs Matrix.

With the obtained Opportunity Cost Matrix it is easy to assess the highest possible regret the DM might feel if each strategy is chosen. That regret will correspond to the highest possible Opportunity Cost of each strategy as illustrated in Table 1.14.

Table 1.14: Opportunity Cost Matrix

		Events, States of Nature or Attributes				**Regret Criterion**
		θ_1	θ_2	...	θ_m	
Strategies or Alternatives	E_1	oc_{11}	oc_{12}	...	oc_{1m}	$\max_j (oc_{1j})$
	E_2	oc_{21}	oc_{22}	...	oc_{2m}	$\max_j (oc_{2j})$
	⋮	⋮	⋮		⋮	⋮
	E_n	oc_{n1}	oc_{n2}	...	oc_{nm}	$\max_j (oc_{nj})$

The Regret Criterion recommends the strategy that minimizes the highest possible regret potentially felt by the DM, *i.e.*:

$$E_k : \min_i (\max_j oc_{ij})) = \max_j (oc_{kj})$$

In other words, the Regret Criterion can be interpreted as the Minimax Criterion being applied to the Opportunity Cost Matrix.

Example

Table 1.15 presents the highest regret for each alternative calculated from the Opportunity Cost Matrix.

Table 1.15: Example's Regret Criterion

		Team's Performances				Regret Criterion
		Very Low	**Low**	**Median**	**High**	
	I	0	**90**	90	23	90
Campaigns	II	62	**165**	147	0	165
	III	25	**87**	15	25	87
	IV	75	0	0	98	98

The recommended Campaign corresponds to the one that minimizes the highest regret: Campaign III with a $\min_j (oc_{ij}) = 87$.

1.2.2   Decision Making in an Risk Context

As previously mentioned in a Decision Making process in a Risk Context (or Under Risk) the occurrence probability of each random event is previously known. And that precious information allows richer and more interesting analyses. SCDM under Risk has been thoroughly and deeply studied. This chapter will present the most well-known and used method for SCDM under Risk problems: the Bayes Criterion.

6. The Bayes[5] Criterion

Also known as the Expected Value Criterion, the Bayes Criterion recommends the strategy that maximizes the Expected conditional value, when this uses an increasing scale. And, obviously, minimizes it if the scale is decreasing. The Expected Values can be calculated because the probability of each event happening is known. It is similar to the Laplace Criterion but considering the actual known probabilities instead of equal probabilities for all events. Table 1.16 illustrates how the Criterion is applied to a generic SCDM problem.

---

[5] Thomas Bayes was an English statistician from the XVI Century. He was responsible for the famous Bayes' Theorem. Bayesian Statistics and the Bayesian interpretation of probability are named as an homage to him.

Table 1.16: Bayes Criterion

		Events, States of Nature or Attributes			Bayes Criterion
		θ_1	θ_2 ...	θ_m	
Probabilities		p_1	p_2 ...	p_m	
	E_1	v_{11}	v_{12} ...	v_{1m}	$\sum_{j=1}^{m} p_j v_{1j}$
Strategies or Alternatives	E_2	v_{21}	v_{22} ...	v_{2m}	$\sum_{j=1}^{m} p_j v_{2j}$
	$\vdots$	$\vdots$	$\vdots$	$\vdots$	...
	E_n	v_{n1}	v_{n2} ...	v_{nm}	$\sum_{j=1}^{m} p_j v_{nj}$

There are two important hypotheses implied in this method that are sometimes unnoticed when the Bayes Criterion is used. On the one hand the method assumes that the DM is Risk neutral, meaning that the DM has an indifferent attitude when facing decisions with different levels of Risk. On the other hand, the method assumes the recommended strategy can be used repeatedly for several times. Only in this situation the problem's expected outcome will be approximated to the one given by the Criterion. If the problem involves a single decision that is taken only once then the strategy outcome can be very different from the Expected value proposed by Bayes Criterion.

Also, bear in mind that the events' probabilities may not be totally independent from the chosen strategy, and in some situations each strategy will imply a different event's probability function.

Example

Table 1.17 adds the National Team's Performance distribution to the example. The expected values for each strategy are:

$$E(I) = 0.1 \times 100 + 0.3 \times 100 + 0.4 \times 100 + 0.2 \times 100$$
$$E(II) = 0.1 \times 38 + 0.3 \times 25 + 0.4 \times 43 + 0.2 \times 123 = 53.1$$
$$E(III) = 0.1 \times 75 + 0.3 \times 103 + 0.4 \times 175 + 0.2 \times 98 = 128$$
$$E(IV) = 0.1 \times 25 + 0.3 \times 190 + 0.4 \times 190 + 0.2 \times 25 = 140.5$$

And the recommended campaign is IV because it is the one with the highest Expected impact.

Using this Criterion while ignoring the implied assumptions may lead to unrealistic, or at least not suitable, recommendations. Also, bear in mind that the events' probabilities may not be totally independent from the chosen strategy, and in some situations each strategy will imply a different events' probability function.

Table 1.17: Example – Bayes Criterion

		Team's Performances				Bayes Criterion
		Very Low	Low	Median	High	
Probabilities		0.1	0.3	0.4	0.2	
Campaigns	I	100	100	100	100	100
	II	38	25	43	123	53.1
	III	75	103	175	98	128
	IV	25	190	190	25	140.5

Consider the following example, adapted from (Goodwin and Wright 2014).

Example

Table 1.18 presents the Pay-offs matrix of a SCDM problem. The matrix includes the conditional profits depending on the success of the launching operation of a new car model, and the probability of the launching operations are a Failure; a Partial Success: or a Full Success.

Table 1.18: Example – Bayes Criterion

	θ_1	θ_2	θ_3
Model 1: Profit	−1	0	3
Probability	0.1	0.1	0.8
Model 2: Profit	−6	1	10
Probability	0.3	0.1	0.6

Note that this is a one time decision process because a car model can only be launched once.

The Expected Profits for each Model is:

$$E(Model\ 1) = 0.1 \times (-1) + 0.1 \times 0 + 0.8 \times 3 = 2.3$$

$$E(Model\ 2) = 0.3 \times (-6) + 0.1 \times 1 + 0.6 \times 10 = 4.3$$

Obviously, Model 2 has the highest Expected Profit and therefore, according to Bayes Criterion, this is the recommended alternative. However, Model 2's launching may fail with a probability that is the triple of the corresponding one for Model 1. Additionally, if the launching fails the potential loss will be six times higher than the less probable loss with Model 1' loss. All things considered, it all boils down to the DM's attitude when facing high risk choices. One way to ameliorate this method's drawbacks is to include the DM's Risk attraction or aversion by resorting to Utility Functions that will be explored in the next Chapter.

1.3 Final Remarks

This chapter introduced the simplest notions of Decision Theory and the most elemental approaches for solving Single Criterion Decision Making problems in a context of Uncertainty and in a context of Risk. Its main goal is to provide the readers with the language and some beginner's tools that will follow them through the rest of this book.

1.4 Proposed Exercises

1. The municipal garden of Vila Nova de Xiripiti will be the target of a huge project of improvement. The budget is already approved in the municipal parliament. However, the garden's renovation works will leave only a few funds for the garden's maintenance. There are three alternative projects for the garden remodeling which have similar working costs, therefore the city councilor responsible for choosing the project will have to carefully consider the projects' maintenance costs. Unfortunately, these costs wildly vary with the humidity level during the next summertime in Vila Nova de Xiripiti. Table 1.19 presents the estimated maintenance costs of each alternative project depending on next year's forecast humidity during the summer.

Table 1.19: Projects' Next Year's Maintenance Costs

Project	Very Dry	Dry	Regular	Humid
		Humidity		
Y	210	130	80	10
X	140	130	120	170
Z	140	140	140	140

(a) The city councilor's optimism level is not known. Therefore propose the most suitable project for all types of Decision Maker, assuming the councilor wants to minimize the garden's expected maintenance cost for next year.

(b) The city meteorologists predict that next summer will be Very Dry, Dry, Regular and Humid with the probabilities, respectively: 20%, 35%, 40% and 5%. What project will you recommend? Comment.

2. The "Adorable Terrors" is a kindergarten and its Director is planning a pedagogic field trip. There are four possible alternatives: (A) the Water Farm, (B) the Botanical Garden, (C) the Children Museum and (D) the "Desperados Shopping Center".

Unfortunately, there are still no weather predictions for the trip day. Table 1.20 presents estimates for the number of parents' protests depending on the weather during the trip day and the chosen destination.

Table 1.20: Expected Complaints for each Possible Weather Condition

	Weather Conditions		
Alternative	Good	Unstable	Bad
Water Farm (A)	20	40	120
Botanical Garden (B)	0	30	80
Children Museum (C)	30	80	20
"Desperados SC" (D)	50	50	50

It's a known fact that the school director is a little pessimistic, but the president of the Parents Association is very optimistic. Is there an alternative that will appease both persons? Justify with a suitable criterion.

3. Auntie Leonarda intends to hire a Personal Assistant for the Party Season. With that in mind she followed four of her nephews and nieces for a couple of days, without them knowing the reasons for her visits. She diligently observed their behaviors during those visits.

In Table 1.21 we are presented the estimated conditional values of "Annoyance Points" she will feel with each potential assistant within different scenarios of media pressure that she will be subjected to during the crazy Party Season.

Table 1.21: Expected "Annoyance Points" for Every Possible Media Pressure Level

	Media Pressure		
Nephew/Niece	High	Average	Low
Urânia	40	15	0
Vitorino	15	10	10
Ximena	15	15	20
Zumélia	13	13	13

(a) Without making any calculations, point out, justifying, which nephew will never be chosen by Auntie Leonarda independently of her optimism level.

(b) "My guts tell me I should choose Vitorino" – Says Auntie Leonarda to herself. Classify Auntie Leonarda, according to her optimism level, justifying adequately, using the Savage Criterion.

(c) Meanwhile, Auntie found out that the probability of suffering a low media pressure during the next party season is solely 14%. She has no idea about how probable it will be for her to have to bear the other pressure levels but usually they are equiprobable. With this new information point out which nephew or niece you recommend based on the Bayes Criterion.

4. Auntie Leonarda will host a small gathering of her nephew Bernardo's little friends. Unfortunately, it will be necessary to keep the brats happy and entertained. And without them breaking her china. So, she is planning on taking the group to spend the day visiting a nice place nearby. **Table 1.22** presents the group's expected average happiness value depending on the chosen place and on the expected weather for the gathering day.

Table 1.22: Group's Average Conditional Happiness Levels

Local	Weather		
	Hot	**Pleasant**	**Rainy**
Alba Beach	100	15	0
Toy Museum	25	25	25
Waterfalls Park	50	80	20

(a) My sweet Bernardo is a kid with a very positive and optimistic attitude. I am sure he will love to spend the day with his friends in the Museum! – thought Auntie. Using a suitable Criterion verify if Bernardo agrees with his Aunt Leonarda.

(b) The weather forecast for the gathering day predicts that the day may be Hot or Pleasant with the same probability which is approximately 14%. Using an adequate Criterion propose an alternative that will minimize Auntie's Expected Regret.

5. The Witch Evarista and her coven sisters, the Witches Zebelina and Hermengarda, came to town to celebrate Samhain. It was a BLAST! Spooking and scaring the mundanes. But, unfortunately they lost the track of time and missed their ride back home in the Murky Mire. Now, the dawn has broken and they have to find a way as discreet as possible to return home ASAP.

There are three magical and hidden portals in the city. They can use one of them to go back. However, the portals are not close to each other and are not infallible. There is always a failure probability when one of the portals is used. And instead of going back home the witches may end up mired in the city. This failure probability varies from portal to portal and is conditioned to the good vibes that were generated in town by the Samhain festivities. Unfortunately, time is running out! Halloween's magical effect is fading and the three women can only try one of the portals before they are deactivated, locking the witches in the mundanes' world for 364 days.

Table 1.23 Presents the usual probabilities of portal failures depending on the generated good vibes.

Without going back to the magical world the witches do not have the means to discover what kind of vibes were generated during last night's celebrations. Nevertheless, Evarista is feeling quite optimistic. And Zebelina's intuition is pointing them to the University's portal.

Table 1.23: Failure Probability Depending on the Generated Vibes

Portal	Samhain Vibes		
	Crystalline	Green	Spooky
By the River	5%	7%	8%
City Park	10%	1%	4%
University Campus	3%	8%	2%

Use Savage Criterion to assess if the two witches are in sync or if Hermengarda, the eldest in the coven, will have to break the tie between Evarista and Zebelina's opinion.

6. Jimmy Long, the IMI16 Agent 700, is investigating an evil plan for World Domination. Jimmy knows the plan is being orchestrated by this mysterious person known simply as X. 700 already discovered X is one of his many archenemies and where the villain's hidden lair is. However, Agent Long is not exactly sure which of his most powerful enemies is actually X.

Now he has to choose the vehicle he will use to infiltrate X's lair. Unfortunately the probability of getting caught depends on the vehicle he chooses and on the real identity of the villainous X.

Table 1.24 presents the probability of Jimmy Long being caught infiltrating the Villain's lair according to the real identity of X and depending on the chosen vehicle.

Table 1.24: Probabilities of being Caught Depending on the Actual Identity of the Villainous X

Vehicle	Actual identity of the Villain currently known as X			
	Madame Aracne	Lord Boltar	Mr. Cranium	The Dormouse
Ghost Car	100	0	90	75
Silent Heli	60	60	60	60
Stealth Bike	80	90	40	75

(a) It was not possible to estimate what are the odds of X being each of Jimmy's arch-enemies. Nevertheless, 700 is inclined to take the Stealth Bike. His boss, agent O, thinks he is being too pessimistic, but Jimmy disagrees. Classify Agent Jimmy Long's optimism level using a suitable Criterion.

(b) Agent 800 is Jimmy's back-up. If something goes wrong, she will go forward with the infiltration mission. She is known to be an extremely pessimistic agent and she really hates to regret choosing a wrong decision. Use the Opportunity Cost Criterion to recommend a vehicle to Agent 800.

(c) Agent O discovered "Madame Aracne", "Lord Boltar" and "Mr. Cranium" have equal probabilities of being X. The Dormouse has a probability of being X that is less than 10%. With this additional information what vehicle do you recommend?

7. Two geologists want to collect samples from an extremely rare mineral that can only be found in four places: the Aramutapol volcano (**A**), the sandy dunes of Bong-Yan beaches (**B**), a deep grotto in the Cliffhanger caves (**C**) and finally, the Dora-Dora Atoll (**D**). Unfortunately they only have the budget to fund just an expedition to one of the aforementioned places.

Sadly, the humidity levels in those places is currently unknown although it is a sure thing that the humidity will affect the expedition costs.

Table 1.25 presents the expedition costs, depending on the humidity level, for each possible destination.

Table 1.25: Expedition Costs Conditioned by the Humidity Level in the Expedition's Destiny

Local	Humidity		
	Low	**Average**	**High**
Aramutapol volcano (A)	80	60	20
Bong-Yan beaches (B)	0	30	80
Cliffhanger caves (C)	50	50	50
Dora-Dora atoll (D)	20	40	120

(a) One of the geologists is a little bit pessimistic and the other is very optimistic. Investigate, using a suitable Criterion, if it is possible to find a place that will please both geologists.

(b) Consider that the probability function for each place's humidity level was correctly estimated and presented in Table 1.26.

Table 1.26: Probability Function for the Expedition's Destinies' Humidity Level

Local	Humidity		
	Low	**Average**	**High**
(A) and (C)	0.15	0.25	0.6
(B) and (D)	0.4	0.35	0.25

In these circumstances what local would you recommend? Justify with a suitable Criterion.

References

Goodwin, P., and G. Wright, G. (2014). *Decision Analysis for Management Judgment*, 5th Edition. New York: John Wiley & Sons Inc.

Knight, F. (1921). *Risk, Uncertainty, and Profit.* New York: Dover Publications Inc.

Sequential Decisions and Introduction to Utility Theory

2.1 Introduction

The previous chapter addressed Single Criterion Decision Making problems and in these types of problems the decision made neither has an impact in future decisions nor is it, supposedly, influenced by previous decisions. But in reality, decisions are often made chronologically and the strategy that is adopted now will sometimes affect not only the available strategies for future decisions, but also the possible but uncontrollable factors that might influence those future decisions. The described situation is a Sequential Decision Making problem and the easiest way to account for all possible conditional scenarios and outcomes is by using a graphic representation of the problem known as the Decision Tree.

Decision Trees are also frequently used to model Utility functions. These functions try to emulate the DM's behavior towards Risk and their use allows the incorporation of DM's Risk attitude in any Decision Theory problem.

This chapter is divided into two major parts: Section 2.2 focuses on Sequential Decision Making and 2.3 presents an introduction to Utility Theory. Sub-sections 2.2.1, 2.2.2, 2.2.3 present how to represent a Sequential Decision Making problem as a Decision Tree diagram and how it can be used to solve the problem. This Section ends with some additional remarks in Sub-section 2.2.4. Sub-sections 2.3.1 and 2.3.2 present two different methods for obtaining an Utility Function from a Decision Making problem and Sub-section 2.3.3 describes the different types of Utility Functions and how they relate with DM's attitude towards Risk. The Section ends with some final remarks. The last Section of this chapter proposes a set of exercises covering essentially Sequential Decision Making.

2.2 Sequential Decision Making and Decision Trees

Sequential Decision Making, as previously mentioned, happens when a decision that has to be made now will affect the availability of additional future decisions, or the available strategies for these future decisions or, even, the random events that might condition the possible outcomes of those decisions. The problem usually involves several sub-problems happening chronologically. The relations between those possible sequential decisions makes the problem complex enough that the use of a graphic representation is not just helpful but fundamental to fully understand all the plausible scenarios. Let's consider an illustrative example.

Example

In Serpa[1] County Hospital, a CT scan machine is malfunctioning. The hospital's daily cost for having the machine out of order is 3,200€. The Radiology Director may opt for one of three alternatives:

- Send the malfunctioning machine to the seller. The maintenance contract ensures the device will be collected, repaired and transported without any additional costs. However, the seller will not be responsible for the out of use costs during the repair time. The seller estimated a repairing time of 10 days that includes the transportation time. Additionally, if somebody else tries to repair the device then the maintenance contract is void and the seller will no longer be an option.
- Assign the equipment repair to an expert technician. The usual repair fee for this equipment is 20,000€ and the technician estimated that the time needed for the repair process is 4 days, with a probability of 70%. Although, with luck, the repair can be done in just 2 days. Out of use costs will be supported by the Hospital.
- One of the resident technicians has the know-how to repair the equipment. With a probability of 50% the equipment will be repaired in 5 days if the repair is dealt by the resident technician. On the other hand, if the equipment is not repaired in 5 days the Radiology Director will have to decide if it is better to call the expert technician or to keep the resident technician repairing the CT scan device. In that case, the resident technician will take two additional days and there is a probability of 25% that the equipment will be repaired. If the resident technician's final try fails then the Director will have to call the expert technician with the costs and probability distribution described above.

What strategy should be recommended to the Radiology Director?

2.2.1   Drawing the Decision Tree

This problem can be represented as a Decision Tree, a diagram that represents the chronological relations between all relevant events and their outcomes. There are two different types of events: Decision events and Chance events. In a Decision event the outcomes are deterministic and the DM will choose the most

---

[1] Portuguese town belonging to Beja District. It is located closer to the Spanish border than to Lisbon, Portugal's Capital. Hometown of one of this book authors.

advantageous one. In a Chance event the outcomes are random and there is a probability function defining the randomness. In a Decision Tree diagram, the Decision events are usually represented by a square and the Chance events are represented by a circle. An example of both can be observed in Fig. 2.1.

The first node, also known as the Decision Tree's Root, is always a Decision Node and represents the Decision that the DM has to make now, knowing what the future possible outcomes of that decision will be, but ignoring what actually is going to happen.

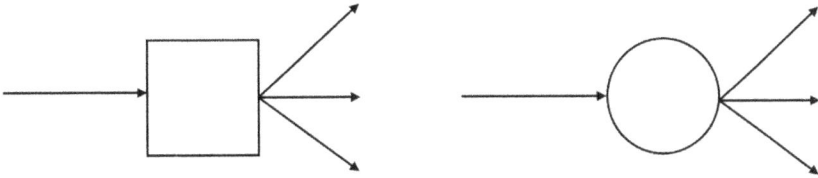

Figure 2.1: Decision Node (left) and Chance Node (right)

The possible scenarios in the end of the tree are also known as Leaves. They can also be represented as nodes in the tree, by triangles, but usually that is not necessary. All nodes are connected by arcs, also known as Tree branches, which represent the chronological relations between two consecutive nodes.

Example

Figure 2.2 shows the Decision Tree diagram that represents the Example's problem. There are nine different possible scenarios, represented by the nine leaves.

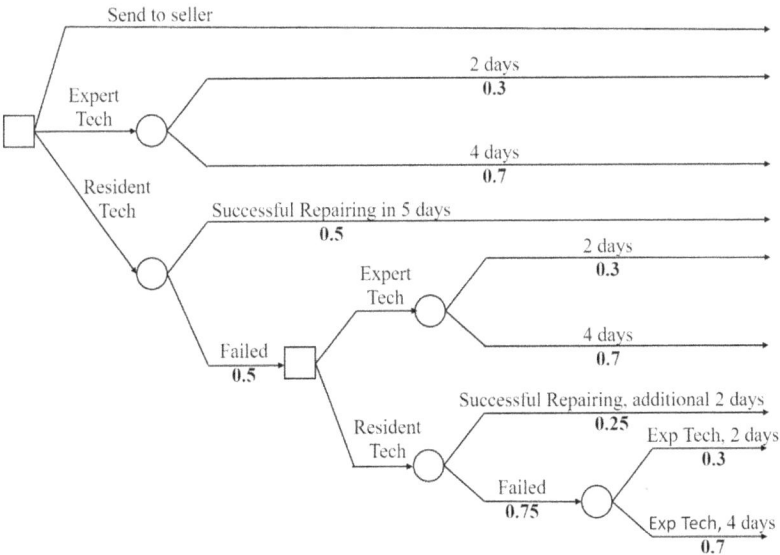

Figure 2.2: Example's Decision Tree

Although it is not necessary, in more complex trees it is usual to merge two consecutive nodes that are from the same type. For two Decision Nodes it is just a matter of merging the available decisions, but for two Chance Nodes it will be necessary to recalculate the probabilities. Fig. 2.3 illustrates this last situation in the example's Decision Tree.

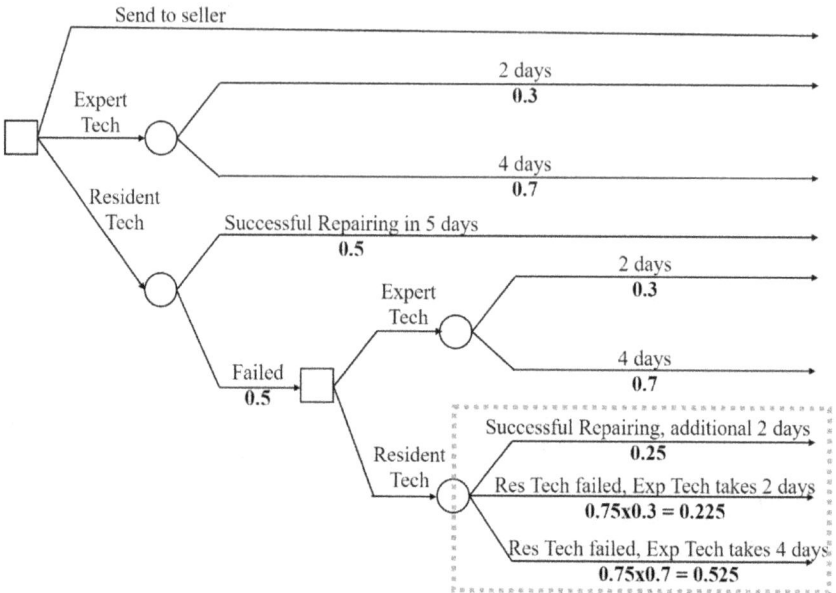

Figure 2.3: Example's Decision Tree with merged Decision Nodes

2.2.2   Assessing the Leaves

As was mentioned before every leaf corresponds to a possible future scenario that can be reached through a succession of decisions and random events. And in every problem there is a function, like Profit, Cost, Time, Distance, or Satisfaction, that is going to be optimized and that allows every scenario/tree leaf to be assessed.

Example:

In the example the objective is to find the decision strategy that will minimize the total cost of each scenario. In the first scenario the equipment will be sent to the seller. The repairing cost will be null, but the device will be out of use for 10 days, corresponding to a total cost of $10 \times 3{,}200 = 3{,}200€$.

In the second leaf the equipment will be repaired by the expert technician (repairing cost of 20,000€) and will be non-operational for 2 days ($2 \times 3{,}200$) for a total cost of $20{,}000 + 6{,}400 = 26{,}400€$.

In Node A_5 there are three possible outcomes, the last three scenarios in the Tree:

- A successful repairing made by the resident technician, with a probability of 0.25 and a cost of $3{,}200€ \times 7$ days $= 22{,}400€$;

- A failed repair from the resident technician, after 7 days, which lead to the repair work from the expert technician, that lasted 2 days, with a probability of 0.225 and a cost of 20,000€ from the fee and 3,200€×9 days, corresponding to a total cost of 48,800€;
- Finally, a failed repairing from the resident technician, after 7 days, which lead to a repairing from the expert technician, that lasted 4 days, with a probability of 0.225 and a cost of 20,000€ from the fee and 3,200€×11 days, corresponding to a total cost of 55,200€.

Figure 2.4 presents the assessment of the nine scenarios. The Nodes were also denominated from A_0 to A_5 so they can be easily addressed in the rest of the Section.

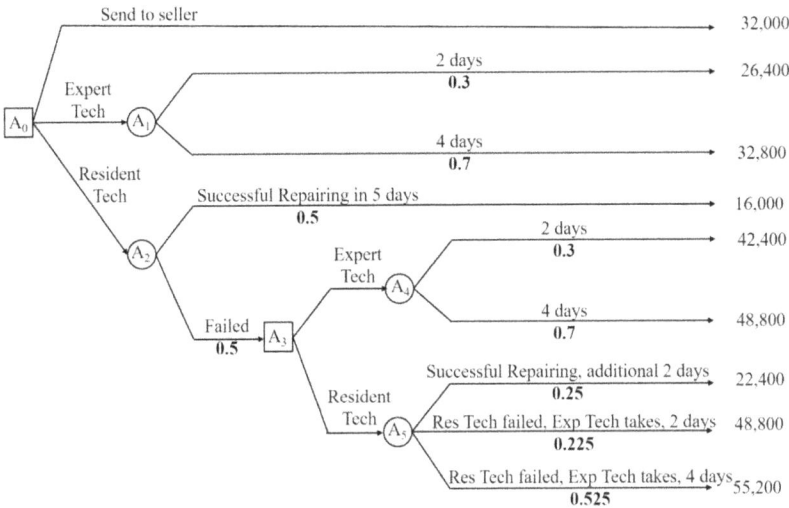

Figure 2.4: Example's Decision Tree with Leaves' Assessment

2.2.3 Assessing the Nodes

Now that the Tree Leaves are assessed the Decision Tree will be analyzed backwards, from the Leaves to the Root, going back in time.

Every Chance Node will be evaluated with the Expected value of its random outcomes, using the adequate distribution.

Example

In Node A_5 the possible outcomes have the following values and respective probabilities:

- 22,400 € (0.25)
- 48,800 € (0.225)
- 52,200 € (0.525)

Therefore the Expected value associated with node A_5 will be:

$$0.25 \times 22{,}400 + 0.225 \times 48{,}800 + 0.525 \times 52{,}200 = 45{,}560 \ €$$

In Node A_4 the possible outcomes have the following values and respective probabilities:

- 42,400 € (0.3)
- 48,800 € (0.7)

Therefore the Expected value associated with node A_4 will be:

$$0.3 \times 42{,}400 + 0.7 \times 48{,}800 = 46{,}880 \ €$$

For the Decision Nodes, the DM will choose the most interesting outcome, considering the expected values on each available outcome.

Example

In Node A_3 the Radiology Director has two available options: calling the expert technician, with an expected cost of 46,880€ (Node A_4) or letting the resident technician try again, with an expected cost of 45,560€ (Node A_5).

The Director will choose the lowest cost and, therefore, the Expected value associated with node A_3 will be *min* (46,880; 45,560) = 45,560 €.

The Expected Costs for all Nodes can be seen in Fig. 2.5.

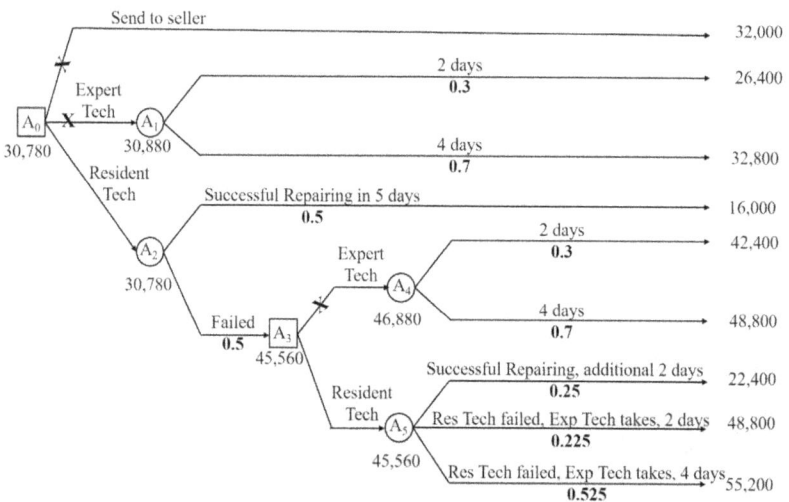

Figure 2.5: Example Solution

The recommended strategy is the one that optimizes the expected value from the available outcomes spanning out from the Root node.

That is the only absolute recommendation that can be made. However, some conditional recommendations can be included in the strategy, describing the succession of most adequate choices in the various Decision Nodes, conditioned by the random events that led to them, until a path from the Tree Root to the optimal Leaf is described. Nevertheless, it must be clear to the DM that it is not possible to ensure that the optimal scenario is reached.

Example

The Director should let the resident technician take a shot at repairing the CT scan device because that option is the one that has the lowest Expected Cost. It is also possible to inform the Director that if the resident technician is not able to repair the equipment in 5 days then the best option is to let him keep trying to finish the task at hand.

2.2.4   Some Comments about Sequential Decisions

This method assumes the DM is Risk Neutral. That means the recommended strategy may not be the most adequate one if the DM is somewhat a Risk Taker or Risk Adverse. One way to ameliorate this drawback is to resort to a Utility Function that models the DM's attitude towards Risk. This will be explored in Section 2.3.

Two additional questions:

- What happens if there is no statistical information relatively to each Chance Node? Then the problem is said to be Under Uncertainty and the methods described in the previous Chapter can be used successively in each Decision Node.
- And what happens if the Leaves' assessments are not deterministic but stochastic? The method can still be used for the average assessments but that can be an approach too simplistic and unrealistic. The most interesting approach, although potentially time-consuming, is a Monte-Carlo Simulation model.

2.3   Utility Theory

Utility Theory was developed by Morgenstern and von Neumann and its goal is to model a DM's attitude when facing different Risk levels. Each DM has an Utility Function that reflects the DM's aversion or tolerance to risks. This function may change depending on the circumstances and important events the DM experiences during his or her Life, although this function is usually stable and unchanged for relatively long periods of time.

A myriad of methods for designing an Utility Function can be found in the literature. In this section we will show two simple and often used methods.

2.3.1   Probability Equivalence Method

This method proposes successive choices of the DM that are adjusted according to the obtained answers. These answers will define several points that will lead to the DM's Utility Function. Each choice (also called lottery), proposes the choice between a gain that is certain and a gamble where there is a fixed probability of a higher gain but also a probability of a loss. A lottery can also propose a choice between a loss and a gamble that has a probability of a gain and a probability of a worse loss. These lotteries can be represented by Decision Trees. Figure 2.6 presents an example of a lottery used for finding the Utility of value X. Usually this is applied to monetary values, but the process can be adapted to any kind of values.

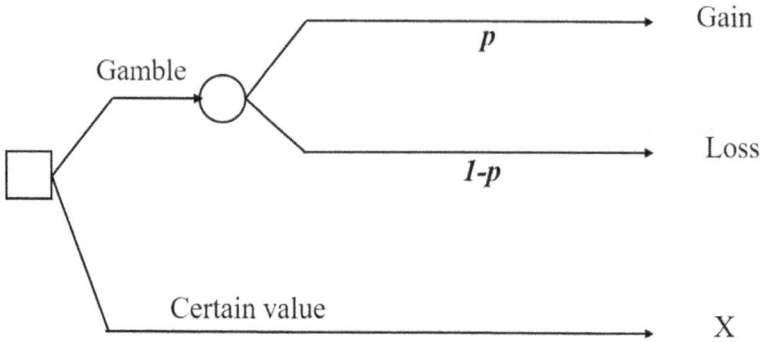

Figure 2.6: Lottery Example.

When faced with this lottery the DM may give three different answers:

1. The DM prefers to receive X, because the gamble is considered too risky;
2. The DM prefers to gamble because the certain gain is considered too low when compared to the potential although random gain;
3. The DM is indifferent between the two choices.

The first answer may indicate a Risk averse DM. To evaluate how adverse the method proposes that a new lottery should be suggested to the DM with an increase of the probability of earning the random gain. And that will be done until the DM becomes indifferent when choosing between X and the gamble.

The second answer may indicate a Risk taker DM. And, again, to assess how far this DM will go in taking risks, a new lottery should be proposed with a lower probability of earning the random gain, successively until the DM becomes indifferent between the gamble and getting X.

The third answer means that, in that lottery, there is an utility equivalence between the gamble and receiving X and $U(X) = p$.

Notice that the Gamble might be between two Gains or two Losses. But in either of these cases one of the outcomes will be obviously much more interesting for the DM.

Example

Consider the following example adapted from (Goodwin and Wright 2014).

A DM has to choose between two different alternatives which have two possible outcomes with the conditional gains presented in Table 2.1.

Recalling Chapter 1, this is a SCDM problem Under Risk and the Bayes Criterion is the most adequate method to solve it:

Table 2.1: Example 2

		θ_1	θ_2
A	Gain	30	11
	Probability	0.6	0.4
B	Gain	60	-10
	Probability	0.5	0.5

- $E(A) = 0.6 \times 30 + 0.4 \times 11 = 22.4$
- $E(B) = 0.5 \times 60 + 0.5 \times (-10) = 25.0$

According to Bayes Criterion, Alternative *B* is the most interesting. Nevertheless, it is the one with the most inherent Risk. Contrary to *B*, Alternative *A* always provides a gain. Although *B* offers the highest gain, also can lead to a loss.

The Probability Equivalence Method assigns a maximum value of Utility, usually 1, to the best possible outcome and zero to the worst, therefore: $U(-10) = 0$ and $U(60) = 1$.

For $U(30)$ the following lottery, in Fig. 2.7, is proposed to the DM, as described above.

The probabilities have to be adjusted until the DM is indifferent between receiving a gain on 30 or gambling. That will be $U(30)$

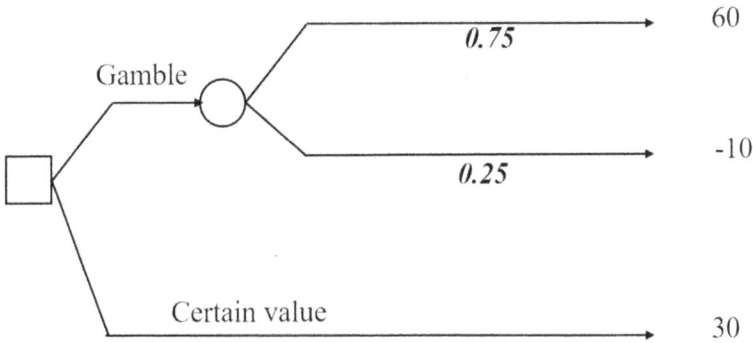

Figure 2.7: Lottery for $U(30)$

The process is repeated for 11, as illustrated in Fig. 2.8.

Due to transitivity $U(11) < U(30)$ therefore the initial value of p, the probability of 60 in the gamble, should be lesser than $U(30)$. The probability presented in Fig. 2.8 is merely illustrative.

Assume that the indifference probabilities, and therefore the Utilities, for the DM are: $U(30) = 0.85$ and $U(11) = 0.60$. That means the Utility Function for the DM is the one presented in Fig. 2.9. But what can be said relatively to this DM's attitude towards Risk? That will be addressed in Section 2.3.3.

Let's calculate the Expected Utilities of example's alternatives:

$$E(U(A)) = 0.6 \times U(30) + 0.4 \times U(11) = 0.6 \times 0.85 + 0.4 \times 0.6 = 0.75$$
$$E(B) = 0.5 \times U(60) + 0.5 \times U(-10) = 0.5 \times 1 + 0.5 \times 0 = 0.5$$

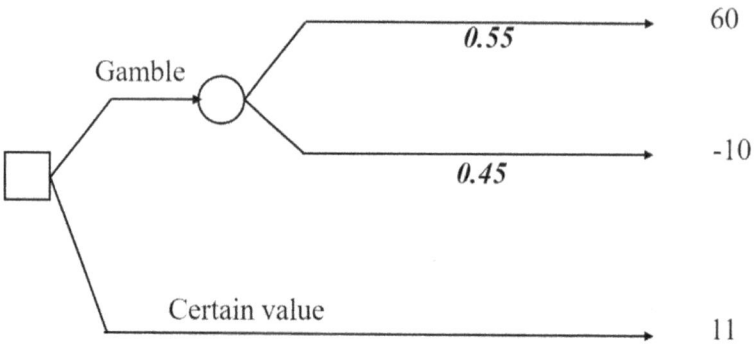

Figure 2.8: Lottery for $U(11)$

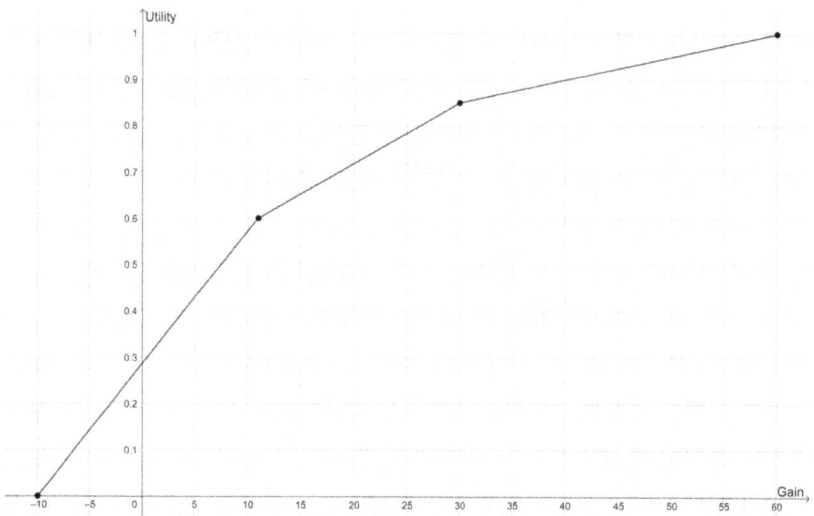

Figure 2.9: Utility Function

And using the DM's Utility function the recommended alternative is A, the one maximizing the Expected Utility.

Although this method is relatively simple, the notion of Lottery between a sure gain and a gamble is not the easiest to explain to a DM. The next method is easier to use but it has its own drawbacks.

2.3.2   Certainty Equivalence Method

Similarly to the previous approach, this method also proposes successive Lotteries, but instead of making the DM choose between a certain gain and the gamble, this method will ask the DM to assign an acceptable price to the gamble. The initial gamble will have equal probabilities for the gain and the loss and this is illustrated in Fig. 2.10.

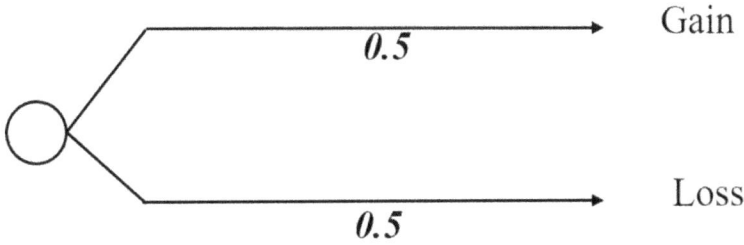

Figure 2.10: Initial Gamble for the Certainty Equivalence Method

The DM has to assign a price to the Gamble that is considered to be a fair one. That price, let's call it X, will be the value that, for the DM, will correspond to $U(X) = 0.5$.

In the next step, the DM is asked to put fair prices to two new Gambles, between X and the Gain and Loss values appearing in the original Gamble. These two new Gambles can be seen in Fig. 2.11.

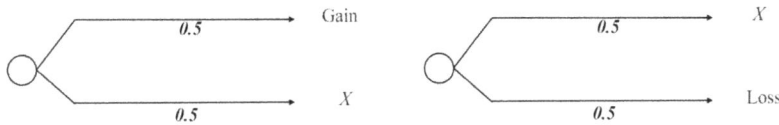

Figure 2.11: Gambles for the Certainty Equivalence Method' Second Step

These two fair prices will correspond to the values with 0.75 and 0.25 Utilities.

The Method can go on until enough intermediate points are obtained.

Example

The first Gamble proposed to the DM to assign a price is represented in Fig. 2.12.

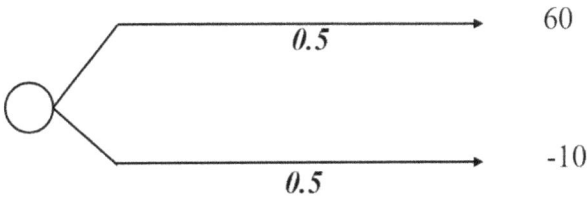

Figure 2.12: Example's First Gamble for Certainty Equivalence Method

Assuming the DM priced this Gamble with 40, meaning that $U(40) = 0.5$, then the next Gambles to be priced will be the ones in Fig. 2.13.

Assuming the DM priced them, respectively, with 55 and 20 that means that the DM's Utility Function is the one presented in Fig. 2.14.

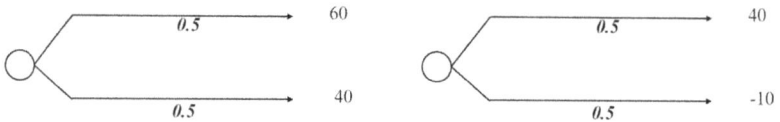

Figure 2.13: Example's Second Step Gambles for the Certainty Equivalence Method

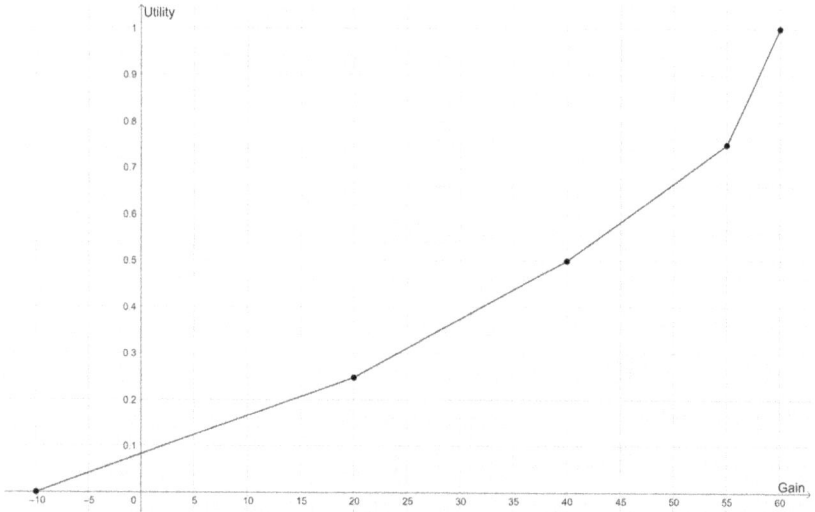

Figure 2.14: Example's Utility Function with Certainty Equivalence Method

This Method is easier to use because usually the DM will feel more comfortable pricing these Gambles rather than choosing between the previous method's lotteries. However, building the DM's Utility Function will make the calculations of the Utilities for the intended values more complicated, because they have to be calculated through linear interpolation.

Example:

In the example it will be necessary to know $U(30)$ and $U(11)$ and neither is among the DM's proposed prices for the Method's Gambles.

Using linear interpolation:

$$20 < 30 < 40 \Rightarrow \frac{U(30)-U(20)}{U(40)-U(20)} = \frac{30-20}{40-20} \Leftrightarrow U(30)$$

$$= U(20) + 0.25 \times 0.5 = 0.375$$

$$-10 < 11 < 20 \Rightarrow \frac{U(11)-U(-10)}{U(20)-U(-10)} = \frac{11-(-10)}{20-(-10)} \Leftrightarrow U(11)$$

$$= U(-10) + 0.25 \times 0.7 = 0.175$$

The Utility functions obtained by the described methods were particularly different solely for illustrative reasons. Although it is common to obtain different Utility Functions when different methods are used the Functions do not usually have such radically different behaviors.

2.3.3 Types of Utility Functions

As was previously shown Utility Functions can have very different shapes because they model different Risk behaviors. Figures 2.15 and 2.16 present the Utility Curves that model the DM's most common attitudes facing Risk.

In Fig. 2.15a, the curve is a typical Utility Function for a Risk Adverse DM. Considering the Probability Equivalence Method, this type of DM only prefers to gamble over a certain gain greater than the average when the probability of winning at the gamble is relatively high. For example, consider a certain gain of 7. The DM will only opt for the gamble if the probability of earning the highest gain is approximately equal or greater than 80%. In opposition, Fig. 2.15b shows the behavior of a Risk taker DM. The gamble is usually preferred even with a moderately high certain gain. For example, for the certain gain of 7, the gamble will only not be preferred if the gaining probability is lower than 50%.

In Fig. 2.16a, it represents the behavior of a Risk neutral DM. The Utility increases linearly with the gain. Finally, in Fig. 2.16b, the typical curve of a DM that is Risk taker until ensuring a gain level and then becoming Risk adverse for higher gain levels. This typically models a situation where the DM has a goal and does not mind taking risks to attain that goal. However, after getting it the DM becomes Risk adverse.

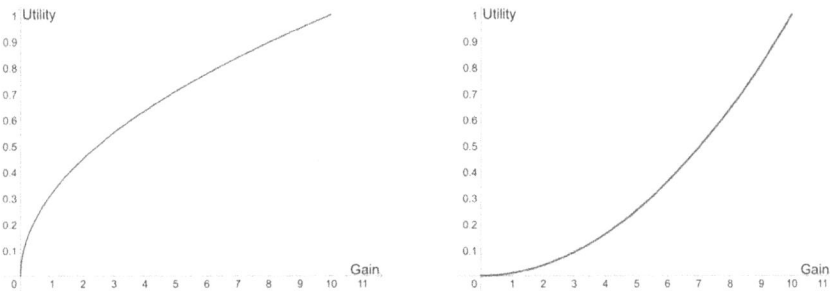

Figure 2.15: Utility Curves
(a) Risk Adverse (left); **(b)** Risk Taker (right)

2.3.4 Final Remarks Concerning Utility Theory

Utility functions try to model a DM's attitude when taking decisions under Risk. However it is fundamental to have in mind that for a specific DM this attitude will change over time, due to Life events and changes in personal perspectives. Additionally a Risk attitude can be very sensitive to the gains values. Usually, for

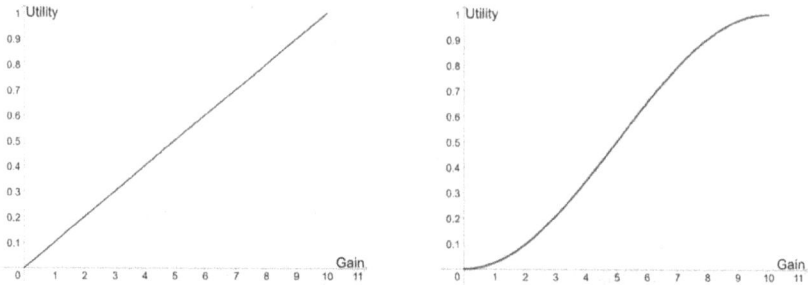

Figure 2.16: Utility Curves
(a) Risk Neutral (left); (b) Both Risk Taker and Adverse (right)

a particular DM, there is usually a range where the Risk attitude will change from Risk taker to Neutral or even Risk Adverse. The "S" curve presented in Fig. 2.16 b) is very common, but depending on the involved values, sometimes only the convex, or the concave part of the function comes into play.

The examples presented in this section described situations with monetary gains and losses, however Utility Functions can model the DM's Risk attitude for other different kinds of Attributes, like distances, time durations, areas, volumes, and so on. Additionally, this section only describes single-dimension Utility Functions, the ones that model the Risk attitude concerning solely one attribute. Actually there are Multi-Dimensional Utility Functions not very different from what will be presented in the next few Chapters.

In spite of its usefulness and of all that was presented in this chapter Utility Theory also bears several drawbacks. Utility is not an easy concept to explain to DM and building an Utility Function requires some answers from the DM to some not so easy to understand questions. If a DM does not answer correctly due to misunderstandings or to boredom, the obtained Utility Function might be totally useless because it will not reflect the DM's true preferences.

Additionally, the use of different methods for building Utility Functions may lead to several paradoxes, from which one of the most famous, is the Allais's Paradox (Allais 1953). Therefore, several authors recommend the use of Utility Functions solely in situations where the DM has a strong Attitude towards Risk and is strong and has a real concern about it. Alternatively, Value Functions, which will be introduced in the next Chapter, are valid methods for modelling DM's preferences. Additional reading concerning Utility Theory can be found in (Goodwin and Wright 2014) and (Rogers 1886).

2.4   Proposed Exercises

1. A truck driver, located in city A, needs to take a load to city B, across the country. To go from A to B there are three alternative roads: R1, R2 and R3. Road

R1 has the most regular traffic flow. The truck driver knows that the journey's duration will be around 15 hours.

The path using road R2 is the shortest. Usually the journey from A to B will only last around 10 hours. Nevertheless, there is an 80% probability of a traffic jam. Fortunately if this happens the truck driver may avoid the jam by choosing one of two secondary roads (R21 and R22) that go around the jammed area.

In case of a traffic jam on R2, alternative R21 has a more inconsistent traffic and the truck driver knows that the journey from A to B can last 12 hours, with a probability of 50%, or 13 hours, with a probability of also 50%. Alternative R22 is shorter than R21, but more dangerous. The journey will have a duration of 11 hours, but there is a 25% probability of a car crash occurring. In that case, the journey will last around 15 hours!!!

The road R3 includes a tunnel that goes through a mountain. Unfortunately, on the news, today, it was announced that the tunnel may be closed, with a probability of 25%. If that happens, the truck driver must travel around the mountain and the journey to B will last 24 hours (!)

If the tunnel is not closed then the journey's duration will depend on the weather. If the weather is fine then the journey will last just 10 hours. But if the weather is foul then the road to the tunnel will be a bit dangerous and the journey will last 13 hours. If the weather is very bad then the total journey will need 16 hours.

The weather forecast predicts a probability of 50% for fine weather, 1/3 for foul weather and 1/6 for very bad weather.

What strategy would you recommend to the truck driver?

2. The Adventurers Guild proposed two missions to Warrick, the warrior: killing Gorumein, the vicious dragon terrorizing the Lands of the Queen; and invading the Black Skull Orcs tribe's fortress.

Warrick visits an Oracle that predicts if he decides to try to remove the dragon from the world of the living he might end up finishing his mission solely lightly injured with a probability of 20%. With a probability of 50% he will defeat the vile reptile but will be suffering from serious injuries. And he might end up not killing the dragon at all and in that case the Guild will register Warrick as a "valiant mission casualty" and gladly send their respectful condolences to the warrior's family and friends.

Concerning the Black Skull fortress, with a 5% probability all is calm and quiet with the Orc tribe but the Guild suspects a new Orc leader is fomenting unrest and inciting the tribe to invade the Lands of the Queen, thus the infiltration mission. If Warrick confirms the invasion is being planned he will have to decide to come back to town and warn the Queen's army or to take a sabotage operation and end the invasion before it blooms. The sabotage operation will succeed with a probability of 65%. But if it fails the Orcs will be furious and use Warrick as a motive for starting the invasion immediately and will catch the Queen's army totally unprepared and the invasion will be an amazing success for the Orcs.

However, if Warrick returns to the Guild with the invasion planning confirmation then the army will have time to prepare for the barbaric invasion and the Orcs' invasion will fail with 70% probability.

Killing Gorumein, the dragon will provide Warrick with 60 satisfaction points (sp). Suffering light injuries correspond to –10 sp and serious injuries to –20 sp. Not surviving implies –100 sp.

Unveiling the invasion is worth 30 sp and successfully sabotaging it awards the warrior with additional 50 sp. A successful invasion corresponds to –60 sp but a failed one is worth 40 sp.

Warrick may choose between one of the two missions or simply escape to another kingdom far, far away from vicious dragons and orcish barbarians, but this last option corresponds to –20 sp.

Which strategy would you recommend to Warrick, the warrior?

3. A SCDM problem generated the Decision Tree presented in Fig. 2.17. The problem is only partially solved. Consider $U \geq 0$.

(a) If $U = 0$ what decision should be recommended in node D1?
(b) Discuss what decision is recommended for all possible values of the parameter U.

4. Consider the Decision Tree presented in Fig. 2.18.

Calculate the value of p that will make the DM indifferent between alternatives A and B.

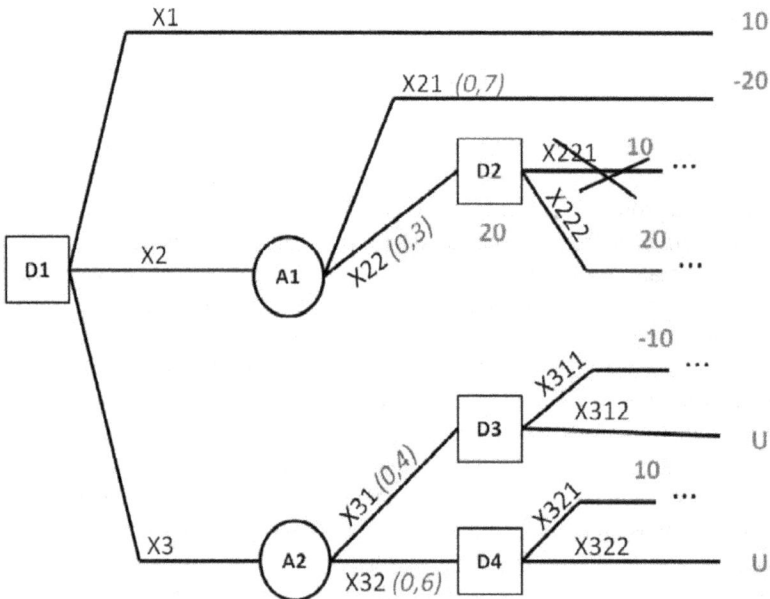

Figure 2.17: Decision Tree

5. Consider the sequential decisions problem represented by the Decision Tree presented in Fig. 2.19. The values on the final nodes correspond to Costs (in m.u.) and *p* represents a probability.

(a) Consider $p = 0.5$. Find what the most adequate decision in Node 1 is.

(b) Find the range of values for *p* that make decision C the most adequate in Node 1.

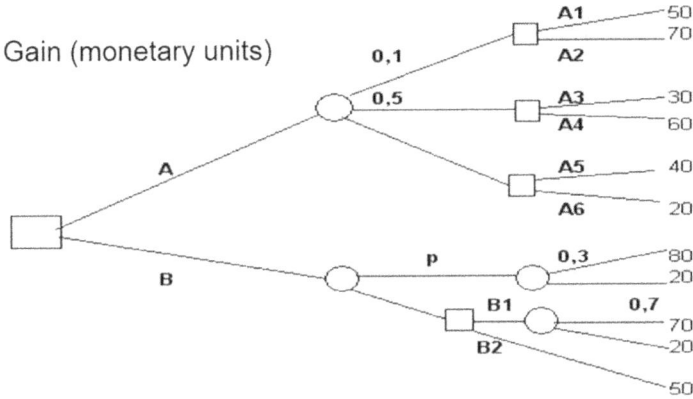

Figure 2.18: Decision Tree

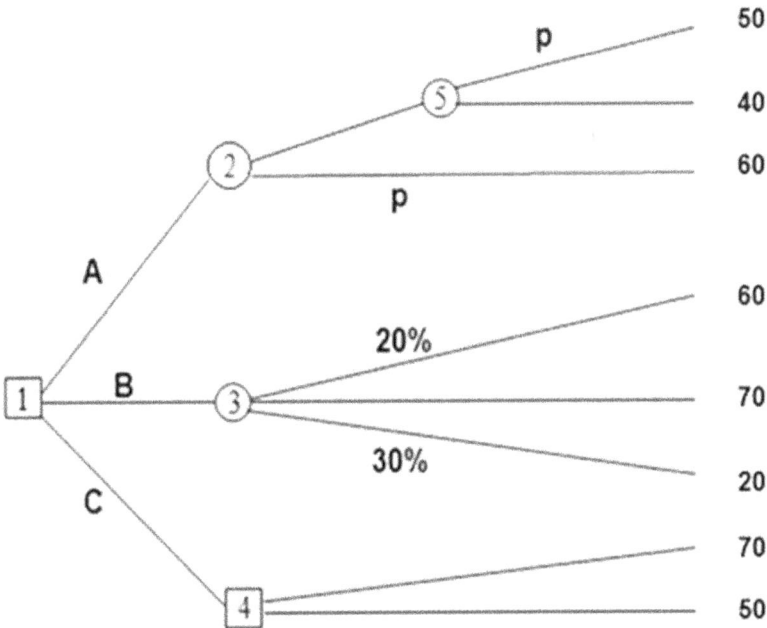

Figure 2.19: Decision Tree

6. The Baroness of Ioa is carefully deciding where to apply her nest-egg because being Nobility is not as a safe bet as it was in the Past. Her godson, Felisberto suggested to her three possibilities: buying treasury bonds; accepting a partnership with Felisberto in his "infallible big monkey business"; and investing in a shares and bonds fund.

The treasury bonds guarantee a profitability rate of 4,5%. The profitability of the shares and bonds fund will depend on the market. Based on past data Felisberto estimated that three possible scenarios may happen during the next two years for the profitability rate:

- A rate of 3.9% with a probability of 0.3;
- A rate of 4.75% with a probability of 0.6;
- A rate of 5%.

Concerning the partnership with Felisberto two situations may arise. If things go bad, the profitability rate will only be 1%. But if things go nicely the Baroness might choose between getting her profit share at the end of first year and deposit in the bank, or wait for the end of the two years.

If she chooses to get her investment early and put it in the bank, she expects to get a profitability rate of 3%. Choosing to keep her mind and money in Felisberto's business for two years may give her a profitability rate of 2% with a probability of 0.25 or a rate of 7% with a probability of 0.75.

The Baroness probed some friends "in the knowing" and estimated that Felisberto's "business" may turn bad with a probability of 0.2.

(a) What strategy would you recommend to the Baroness?
(b) Admit the Baroness' Utility Function for the profitability rate is represented in Fig. 2.20.

   (i) Classify the Baroness' Risk attitude.
   (ii) What strategy would you recommend considering the information given by the Baroness' Utility?

7. The Baroness is visiting Mockistan and she is planning next week's tour to the city Ganxa. The Baroness is hosted in the city of Tarep. She can go from Tarep to Ganxa by bus or by train. The bus trip may last 1, 2 or 3 hours respectively with probabilities of 0.1, 0.4 and 0.5, depending on the military traffic in the area. However, that trip is no fun at all because it crosses long extensions of boring desert.

The train trip crosses the beautiful White Mountains with breathtaking views. But the trip will last much longer. The trip duration from Tarep to Ganxa depends mostly on the weather in the mountains. If it does not rain torrentially the trip can be made up and down the mountains and will last around 5 hours but the Baroness will enjoy the most stunning views. However, if it rains torrentially the mountain crossing will be interrupted and the Baroness will have to change to one of two alternative train lines that go around the mountains. The North line, through Almexunia region makes the Tarep-Ganxa trip in 5 hours. The landscape

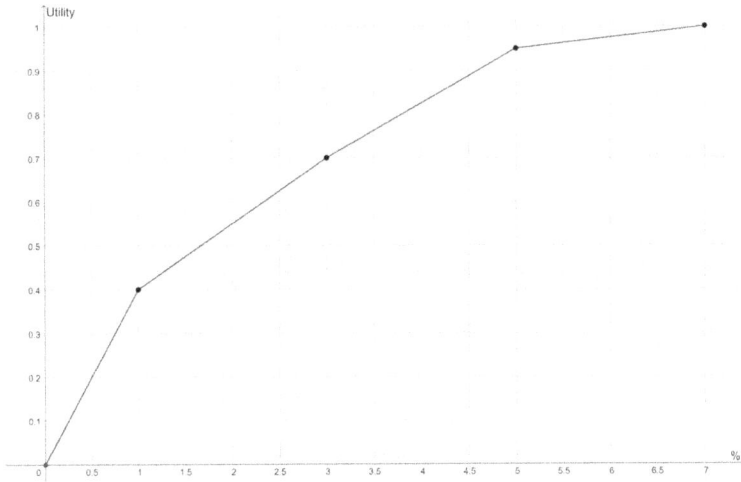

Figure 2.20: Baroness' Utility Function

is still very pleasant, although not grandiose like up in the mountains. The trip using the South line, going through Bustinopolis, will have a duration of 4 hours but the landscape is less interesting, although still much more beautiful that the one in the desert. Unfortunately, there is an additional catch: both North and South lines may be temporarily interrupted by an elephant herd. If this happens it will increase the trip duration by 2 hours, independently of the line being used.

The Mockistan Train Company estimated the following probabilities for next week:

- Torrential rain in White Mountains 40%
- Elephant herd interrupting North line 25%
- Elephant herd interrupting South line 15%

(a) Suggest to the Baroness a plan for a Ganxa trip considering she intends to minimize the expected trip's duration.
(b) How can the Baroness' enjoyment with the trip be incorporated in this problem?

References

Allais, M. (1953). Le Comportement de l'Homme Rationnel devant le Risque: Critique des Postulats et Axiomes de l'Ecole Americaine. *Econometrica*, 21(4), 503. https://doi.org/10.2307/1907921

Goodwin, P. and G. Wright (2014). *Decision Analysis for Management Judgment*, 5th Edition. New York: John Wiley & Sons Inc.

Rogers, A.E. (1886). The theory of utility. *Science*, *ns*-8(193).

Simple Multi-Attribute Rating Technique – SMART

3.1 Introduction

The SMART technique was proposed by Edwards in 1971 (Edwards 1971). It is a simple method that generates robust solutions. Currently, it still is one of the most used multi-criteria methods due to its simplicity. One of its main advantages is the need for a constant dialogue between the Decision Maker (DM) and the analyst, which leads to the DM feeling a direct involvement in the decision process. This chapter will present the methodology and its successors, (SMARTS and SMARTER) (Edwards and Barron 1994) and describe their advantages and some limitations. The methodologies will be explored with the support of an example. This example will be used also in Chapters 4 to 6, with the necessary adaptations to the specifics of the methods introduced here and in the mentioned chapters. Therefore, the example is presented in the Appendix for ease of access.

This chapter is organized as follows. It starts with the presentation of the methodology with its nine-step approach. Then, sections 3.2.1 to 3.2.5 thoroughly address concepts as value trees, alternatives' assessment, value functions, dominance, and weights assignment. How to calculate the alternatives final scores and how robust they are is tackled in sections 3.2.6 and 3.2.7. Robustness is evaluated through sensitivity analysis. Two methods are presented, one of general use and one designed for three criteria problems. The chapter ends with some final remarks and a set of exercises covering all topics.

3.2 The Method

The SMART technique follows a methodology that can be described in nine steps:

1. Identification of the DM.
2. Identification of Alternatives.
3. Identification of DM's points of view that are considered relevant for the decision process and how to assess them (Attributes).
4. Evaluation of all Alternatives according to the Attributes previously defined.

5. Conversion of all Attributes to a reference numerical scale so they can be compared (Criteria).
6. Dominance Analysis.
7. Attribution of weights to each Criterion.
8. Determination of a Weighted Sum for each Alternative using the weights and the Evaluations previously defined.
9. Implementation of a Sensitive Analysis to assess the result's robustness.

The first three steps are common to all Multicriteria Decision methodologies therefore their description and the relevant definitions are presented in the Appendix. In this chapter the authors will focus on the characteristics that are unique to SMART, SMARTS and SMARTER.

Steps 1 and 2 should be very straightforward. Step 3 is usually hand waived in the majority of MC methods but the SMART family considers it rather important because it is the first step incorporating the DM perspective and subjectivity in the process.

3.2.1   Value Trees

When confronted with the question about which points of view are more important for a decision most DM will give some vague and rather generic answers like "the technician must be a nice person" or "I want to hire the best person for this job". These points of view, however relevant for the DM, are not useful for a Decision Process because they are not easily accessed and measured, therefore they can not be considered Attributes. Deciding which Attributes to use may not be a trivial task but creating a Value Tree can be a tremendous help for a less sure DM.

A Value Tree is a hierarchy like structure that begins with general concepts at the top, representing the DM's most important, although vague, concerns, and branches into more specific notions towards the bottom, until it ends with concrete measuring tools that can be used to access the Alternatives, that is, the Attributes.

Consider the example presented in the Appendix. The DM's main concerns were, as mentioned above, "The hired candidate should be the most relatable" and "The hired candidate must be the most suitable for the job". So, this could perfectly be the tree's top levels. Figure 3.1 represents a possible Value Tree for the example problem. Values Trees are built from Top to Bottom during a dialogue with the DM and the main goal is to specify ways of measuring the alternatives' performance concerning the DM's most important points of view and it is the first way of SMART incorporating DM's subjectivity in the decision making process.

Building a Value Tree is obviously a highly subjective process. Nevertheless, according to (Goodwin and Wright 2014) some aspects should be taken in consideration. Value Trees should be:

- **Complete:** All DM's most important concerns should be included.
- **Operational:** The Attributes at the bottom level of each tree's branch should be easily assessed by the DM.

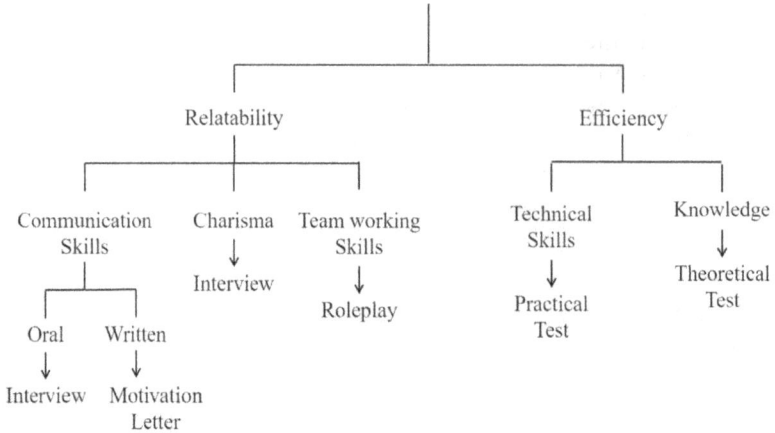

Figure 3.1: Value Tree for Example Problem.

- **Minimal:** Too many Attributes will drag the Analysis and make it too repetitive probably boring the DM.
- **Non Redundant:** It should not include two Attributes that assess similar DM's points of view. A Value Tree with redundancy might lead to the same point of view contributing twice to the final aggregated sum.
- **Decomposable:** The assessment of an Alternative by one Attribute should be independent from the result of the assessment from the other Attributes. If the DM needs to check the Alternative's performance in one Attribute before deciding how to grade it in another then the Value Tree is not decomposable.

It is often necessary to find a compromise between some of the aspects mentioned above. For example, to ensure a Tree's decomposability it may be necessary to increase its size, compromising the minimum size.

3.2.2   Alternatives' Assessment

Now that the Attributes were chosen it is time to assess all the Alternatives' performance in each of them. For some Attributes, like "Distance", "Cost", "Number of visitors per week", for example, that assessment can easily produce a quantitative grade. But for other Attributes, like "Comfort", "Visibility", "Ease of use", obtaining a numerical evaluation is not such a trivial task. In the following, two solutions for these situations will be explored: Direct Rating and Qualitative Scales.

Direct Rating

The Direct Rating technique allows the ranking of all alternatives from an Attribute's perspective without any previous assessment. The first step is to ask the DM to sort all alternatives considering one specific Attribute. Consider the example presented in the Appendix but include another candidate "E" that obtained 6 demerits in the Practical Test and 75.0 % in the theoretical test. The DM has already interviewed the candidates and the Direct Rating is going to be used to assess their performance during the interviews. The DM is now asked to sort the five candidates considering their performance in the interview. After thinking for a while the DM proposes the following ranking:

$$1^{st} – A; 2^{nd} – E; 3^{rd} – D; 4^{th} – B \text{ and } 5^{th} – C$$

The second step is to score the candidate in the last position with zero points. The aim is to score the candidate in the first position with 100 points but respecting some subjective scale that is going to be built according to the DM's preferences.

In the third step, the DM is asked to score the candidate in the second-last position with whatever value the DM feels comfortable using. Assume that candidate B scored 10 points. Figure 3.2 shows the beginning of the rating. This increase in value between the worst and the second to last candidates will be the standard used in the next steps, which will iteratively build the scale from bottom to top.

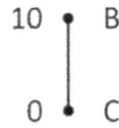

Figure 3.2

Now comes the trickiest steps, in the fourth step, the DM is asked to consider the increase in value from the candidate in the 4th position to the one in 3rd. And, after, to compare it with the corresponding increase between the worst candidate and the one in the 4th position. After some deliberation, the DM considered the increase in value between the 4th and the 3rd candidates to be approximately half of the same increase between the 5th and the 4th, corresponding to an increase of 5 points. Figures 3.3 and 3.4 illustrate the interaction with the DM during this step.

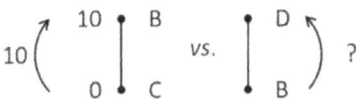

Figure 3.3: Step 4, Question to DM.   **Figure 3.4:** Step 4, DM's Answer.

With this new information the current scale can be updated as can be seen in Fig. 3.5.

Step 4 has to be repeated for the remaining Alternatives, A and E. Assume the increase between the candidates in the 3rd and 2nd positions is the triple of the standard, as is illustrated in Fig. 3.6.

And, finally, for the best candidate, A, the DM considered the increase in value between the candidates in the 2nd and 1st position will also be half of the standard increase (between the 5th and the 4th candidates).

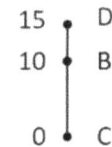

Figure 3.5

The final scale will include all this information and can be seen in Fig. 3.7.

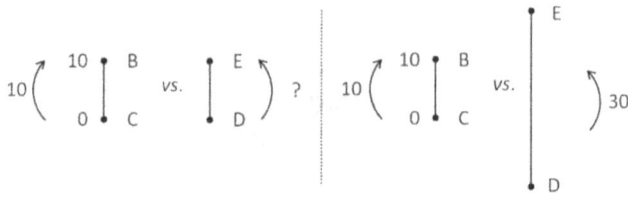

Figure 3.6: Step 4 for Candidate E.

Nevertheless, the obtained scale does not match with what was mentioned in the beginning of this process: the best candidate should get a score of 100. Step 5 will correct this issue. All scores are divided by the current best score and multiplied by 100. This linear transformation will ensure the candidate in the first rank will have a score of 100 and the middle-scale candidates will be scored with the corresponding value between 0 and 100 that verifies the relation between all scores calculated from the DM's inputs.

Table 3.1 compiles the current assessment of all candidates according to the three Attributes. In this Chapter an additional candidate was included (Candidate E) in the example because SMART implements an optional analysis seldom considered in other methodologies and thus this candidate will have a particular relevance in section 3.2.4.

Figure 3.7

Table 3.1: Candidates Scores

		Interview	Demerits in Practical Test	Theoretical Test
	A	100	4	75.4%
	B	20	2	84.6%
Candidates	C	0	3	95.7%
	D	30	6	90.2%
	E	90	6	70.5%

The Alternatives performances according Attribute "Interview" were rated in a 0 to 100 reference scale. This scale will also be applied to the other Attributes. When all performances are rescaled to this reference scale the Attributes become Criteria.

The main disadvantage of using the Direct Rating approach is obvious: if a new Alternative has to be included in the Decision process in a later phase, the Direct Rating technique has to be applied again because the Alternatives' rates are relative.

Qualitative Scales

Sometimes when it is not easy to assess quantitatively an Alternative's performance in an Attribute it is easy to just use a qualitative scale such as "Very Bad", "Bad", "Regular", "Good", "Very Good". The number of adopted categories is mostly irrelevant. The important thing is for the DM to be comfortable when assigning a category to an Alternative based on its performance. Consider the Candidates' performance presented in Table 3.2.

Table 3.2: Candidates Scores using a Qualitative Scale

		Interview	Demerits in Practical Test	Theoretical Test
	A	Very Good	4	75.4 %
	B	Good	2	84.6 %
Candidates	C	Regular	3	95.7 %
	D	Good	6	90.2 %
	E	Very Good	6	70.5%

Now, the Qualitative Scale's categories have to be transformed into numerical scores and that can be done by a process similar to Direct Rating but considering the categories. Assume that the DM assigned the following scores to each alternative:

Very Bad – 0; Bad – 10; Regular – 30; Good – 60; Very Good – 100

Table 3.3 shows the Candidates' performances after the scoring presented above is applied to Attribute "Interview".

Table 3.3: Candidates Scores after Qualitative Scale's Rescaling

		Interview	Demerits in Practical Test	Theoretical Test
	A	100	4	75.4 %
	B	60	2	84.6 %
Candidates	C	30	3	95.7 %
	D	60	6	90.2 %
	E	100	6	70.5%

This approach's main advantage is the introduction of a new Alternative which is easily solved. The DM will solely assign a qualitative category to this new Alternative and its score will be applied automatically because the scoring was created over a generic qualitative scale.

3.2.3   Value Functions

As was mentioned before for some Attributes it is easy to quantify the Alternatives' performances. Nevertheless, maybe the observed numerical values do not have a linear relation with the subjective importance level considered by the DM. For example, consider the candidates' results in the written theoretical test presented in Table 3.3. Candidates A and E have a difference of 5% between their performances. A similar difference can be seen between candidates C and D. Nevertheless the first pair has grades around 70% and the last one around 90%... One can wonder if the DM considers both differences of 5% as equals. Maybe, an increase from 90% to 95% has, subjectively, much more of an impact than an increase from 70% to 75%, because it is harder to get. So, how to incorporate this kind of subjectivity into the Decision process? Value Functions can provide help in this particular matter.

There are two methods that can be used to build an adequate Value Function. The first one assigns a score to each numerical score using the reference scale in a process analogous to the Direct Rating technique. The second process defines a generic function that relates the current scale with the reference scale.

Rating the Numerical Scores Using the Reference Scale

The first method assigns the value of zero to the worst score and 100 to the best one. In the example, consider the written test score.

The worst and best results are, respectively, 70.5 % and 95.7%. So:

$$v(70.5) = 0 \text{ and } v(95.7) = 100$$

The remaining scores, 75.4%, 84.6% and 90.2% are evaluated directly by the DM, not differently from what was done in the Direct Rating technique:

$$v(75.4) = 10, v(84.6) = 50 \text{ and } v(90.2) = 90.$$

Notice that for Attribute "Demerits in the Practical Test" this process will invert the scale's direction. The worst performance corresponds to the highest number of demerits and *vice-versa*:

$$v(6) = 0 \text{ and } v(2) = 100$$

for the worst and the best performances. The remaining:

$$v(4) = 40 \text{ and } v(3) = 80.$$

Similarly to Direct Rating, the inclusion of an additional Alternative will cause the recalculation of all values because the relative position of each Alternative comparatively to the others will change.

Building a Generic Function

In the second method the Value Function does not depend on the Alternative's performances. It assigns an acceptable scale to the Alternatives' performances and demands a new judgement call from the DM which introduces a new level of complexity. On the other hand, because the built Value is independent from the Alternatives' performances, the introduction of an additional Alternative is a simple issue.

The first step is to establish with the DM what the Attribute's scale bounds are. For instance, for the Attribute "Grade in the Written Test", the upper bound is easy to settle: 100 %. For the lower bound the DM has to decide what would be the lowest acceptable grade. After some consideration, the DM decided grades below 70% would lead to the candidate's elimination. This decision corresponds to:

$$v(70) = 0 \text{ and } v(100) = 100.$$

For the intermediate values the DM will have to choose what grade will correspond to the middle point between the best and the worst grade. If the DM has a linear subjectivity the value will be around 85%, but most times it is not, and that value will reveal something about the DM's biases concerning the Attribute. Assume the answer is 90% which means that $v(90) = 50$. Now the process is repeated for two other mid-points: the one between the worst grade and the half-of-the-scale grade and the midpoint between the half-of-the-scale and the best grades. These mid-points correspond, respectively, to the grades with values 25 and 75. Assume these grades were:

$$v(85) = 25 \text{ and } v(92.5) = 75.$$

Figure 3.8 presents the Value Function built from the DM's inputs concerning the acceptable grades in the written theoretical test.

With the Value Function in hand it is time to calculate the value corresponding to the grades obtained by the candidates. Those values will be obtained through linear interpolation. For example, for the best grade, from candidate A, 95.7%:

$$92.5 \leq 95.7 \leq 100 \Leftrightarrow v(92.5) \leq v(95.7) \leq v(100)$$

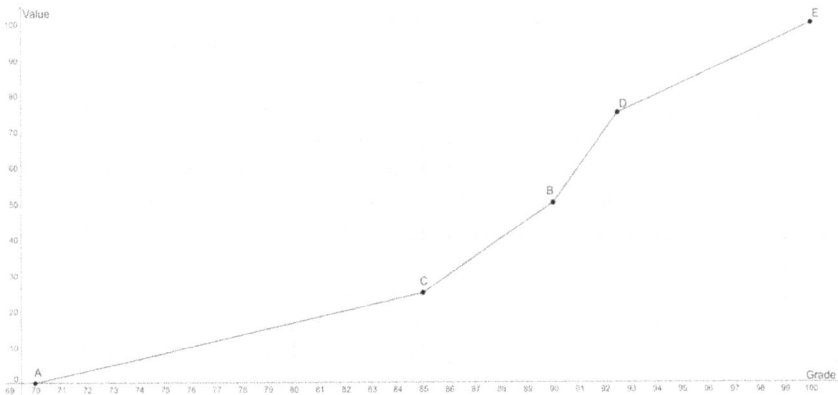

Figure 3.8: Obtained Value Function.

$$\frac{v(95.7)-v(92.5)}{v(100)-v(92.5)} = \frac{95.7-92.5}{100-92.5} \Leftrightarrow v(95.7) = v(92.5) + \frac{95.7-92.5}{100-92.5}$$

$$\times (v(100)-v(92.5)) \Leftrightarrow v(95.7) = 75 + \frac{3.2}{7.5} \times 25 = 85.67$$

Another example, for the grade of 90.2:

$$90 \le 90.2 \le 92.5 \Leftrightarrow v(90) \le v(90.2) \le v(92.5)$$

$$v(90.2) = v(90) + \frac{90.2-90}{92.5-90} \times (v(92.5)-v(90)) = 50 + \frac{0.2}{2.5} \times 25 = 52.0$$

Table 3.4 presents the values of the remaining candidates' performance considering the "Grade in the written test" Attribute.

As was previously mentioned, this method allows an easy insertion of additional Alternatives. Nevertheless, the obtained values are very dependent on the scale bounds, therefore it is quite common to test several different bounds until the DM is comfortable with the resulting values.

Obviously, for Attributes with a decreasing scale, such as the "Number of demerits in the Practical test", this method will also invert the scale direction.

There are alternative methods that allow the incorporation of the DM's subjectivity in the re-scaling process. One of the most common is to use a generic pre-established Value Function. Consider x the performance of a given Alternative according to an Attribute, with $x \in [x_{min}, x_{max}]$. The DM's subjectivity can be incorporated by choosing an adequate value for α in the following Value Functions:

For Attributes with decreasing scales: $v(x) = M \times \left(\dfrac{x_{max} - x}{x_{max} - x_{min}} \right)^{\alpha}$

For Attributes with increasing scales: $v(x) = M \times \left(\dfrac{x - x_{min}}{x_{max} - x_{min}} \right)^{\alpha}$.

Notice that for $\alpha = 1$ the Value Function is a linear function that transforms the interval $[x_{min}, x_{max}]$ into $[0, M]$. Usually, $M = 100$. But it is also common to use $M = 1$, $M = 10$ or $M = 20$.

Table 3.4: Values for the "Grade" Attribute

		Grade in Written Test	Value
	A	75.4 %	9.00
	B	84.6 %	24.33
Candidates	C	95.7 %	85.67
	D	90.2 %	52.00
	E	70.5%	0.83

When all Attributes' scales are converted to the reference scale then an Alternative's performances in different Attributes are comparable. That means that the Attributes were converted into Criteria.

Consider Table 3.5, with a possible conversion of the three Attributes into Criteria. Criterion "Interview" uses the values obtained through the Qualitative scale, and the Criterion "Grade in the Theoretical Test" uses the first method for building Value Functions. How the values for Criterion "Demerits in the Practical Test" is irrelevant for what follows.

Table 3.5: Criteria Table

		Interview	Demerits in Practical Test	Theoretical Test
	A	100	50	10
	B	60	100	50
Candidates	C	30	75	100
	D	60	0	90
	E	100	0	0

Up to this point it was assumed all alternatives were available and that each alternative assessment is a deterministic process. If the assessment of an alternative's performance happens to be stochastic then the use of Utility Functions instead of Value Functions is recommended, especially if the DM is concerned with the underlying risks.

3.2.4   Dominance Analysis

Dominance Analysis allows the removal of Alternatives that will never be able to achieve the top rank in the final ranking and, in certain circumstances, they can be removed from the decision process. Consider a MCD problem with n criteria C_1, C_2, ..., C_n, and $v_i(X)$ the value of alternative X according to Criterion i.

Alternative X is said to be dominated by Alternative Y, or that Y dominates X if and only if:

$v_i(X) \leq v_i(Y)$, for all i $\in \{1, 2, ..., n\}$ and

$\exists k \in \{1, 2, ..., n\}$ where $v_k(X) \leq v_k(Y)$.

For example, from Table 3.5 it is easy to see that:

$v_1(E) = v_2(A) = 100$, $v_2(E) = 0 < v_2(A) = 50$ and $v_3(E) = 0 < v_3(A) = 10$.

Therefore Alternative E is dominated by Alternative A. Alternative E will never be in the 1st rank because will always have a lower rank than Alternative A.

In a MCD problem where the goal is to find the most adequate Alternative for a given DM, only the top rank Alternative really matters. Consequently, all dominated Alternatives can be removed from the process and the final steps of SMART can ignore them. But in other circumstances it is not recommended to do that removal. For instance, if more than one Alternative is to be selected by the DM the removal of a dominated Alternative could lead to the selection of the first and third most interesting options, because the dominated Alternative might have ended in the 2nd rank of the final ranking if it had not been removed. Moreover, in some situations it is mandatory to publish the final ranking of all Alternatives. In this case, every Dominated Alternative has to be kept in the decision process until the end.

In the example, because only one candidate will be chosen, Alternative E might be removed from the decision process. But, if the Science Lab was being funded to the Government than local legislation would probably demand the publication of the final ranking including all candidates and, therefore, Alternative E must be kept in the process for the final steps.

3.2.5   Assigning Weights to the Criteria – Swing Weight

After the procedures presented in the previous sections all Alternatives' performances according to all Criteria will be valued in a reference increasing scale where the lowest and highest possible values for each Criteria corresponds to 0 and 100, respectively. To get the final classification of one Alternative is necessary to combine the obtained values in each Criterion. But how to combine them? If all Criteria had the same relative importance for the DM a simple average could be a good way to integrate all values in a single indicator. Unfortunately, that equal importance seldom happens. So, in this step it will be necessary to translate the DM' opinion concerning the Criteria relative importance into Weights that will be used to calculate a weighted sum for each Alternative.

Originally, the SMART method did not include the Swing Weight technique. According to (Edwards and Barron 1994) this method for assigning weights to the criteria was created by a non-identified analyst working at Decisions and Designs, Inc. During the 70's Swing Weights was incorporated in SMARTS (SMART with Swings) but its authors did not know who they should credit and nowadays most people believe Swing Weights was created by (von Winterfeldt and Edwards 1986).

How does Swing Weight work? The technique follows some easy steps presented below:

1. Sort all criteria from the most important for the DM to the least;
2. Assign a weight of 1 (one) to the most important criterion;
3. For each of the other criteria, C_k ask the DM to compare two fictional alternatives with the following characteristics:

    Alternative X – Has a score of 100 in the most important criterion and 0 in the others;

    Alternative X' – Has a score of 100 in criterion C_k and 0 in the others.

4. The DM will assign what is the percentage of interest that Alternative X' has relatively to Alternative X. That percentage will correspond to the weight of the respective Criterion C_k.

   Steps 3 and 4 are repeated for all Criteria other than the most important.

5. Standardize the weights so their sum equals to 1.

The expression "Swing" comes from the comparison of the two fictitious Alternatives, where the weights swing from 0 to 1 in two different Criteria.

Let's apply this technique to the example. The DM considers C_1 – "Interview result" to be the most important criterion, followed by C_2 – "Demerits in the practical test" and finally, C_3 – "Grade in Written Theoretical Test". Therefore, the weight for C_1 is $w_1 = 1$. For C_2, the two fictional alternatives are:

$$X: \begin{matrix} C_1 & C_2 & C_3 \\ 100 & 0 & 0 \end{matrix} \qquad X': \begin{matrix} C_1 & C_2 & C_3 \\ 0 & 100 & 0 \end{matrix}$$

Assume that the DM believes that X' is just 80% interesting when compared to X. That means $w_2 = 0.8$. Now, for C_2, the two fictional alternatives are:

$$X: \begin{matrix} C_1 & C_2 & C_3 \\ 100 & 0 & 0 \end{matrix} \qquad X'': \begin{matrix} C_1 & C_2 & C_3 \\ 0 & 0 & 100 \end{matrix}$$

In this case, the DM believes that X'' has solely 20% of interest when compared to X. So, $w_3 = 0.2$. However, $w_1 + w_2 + w_3 = 1 + 0.8 + 0.2 = 2$. Standardizing the weights:

$$w_1 = \frac{1}{2} = 0.5, \quad w_2 = \frac{0.8}{2} = 0.4, \quad w_3 = \frac{0.2}{2} = 0.1$$

And here are the weights for the three criteria. Now, it is possible to calculate the weighted sum for all alternatives.

3.2.6   Calculate the Weighted Sum for All Alternatives

Table 3.6 presents all relevant information calculated in the previous sections and includes the weighted sum for all Alternatives. For the sake of completeness, the

Table 3.6: Weighted Sums for all Alternatives

		Interview	Demerits in Practical Test	Theoretical Test	Weighted Sum
	Weights	0.5	0.4	0.1	-
	A	100	50	10	71.0
	B	60	100	50	75.0
Candidates	C	30	75	100	55.0
	D	60	0	90	39.0
	E	100	0	0	50.0

weighted sum for Alternative C, for example, is $30\,w_1 + 75\,w_2 + 100\,w_3 = 30 \times 0.5 + 75 \times 0.4 + 100 \times 0.1 = 55$.

As was expected, Alternative E, dominated by A, has a weighted sum lesser than A. Nevertheless, it is not the Alternative with the lowest sum. That reflects the considerations explored in section 3.2.4 concerning the removal of dominated Alternatives.

From Table 3.6 it is finally possible to rank the five Alternatives, using the weighted sum:

1st – B (75); 2nd – A (71); 3rd – C (55); 4th – E (50); 5th – D (39).

But a question arises: Will the ranking still be the same if the weights change? And how much can they change without causing a radical change in the ranks? These questions address the ranking's robustness and these questions will be answered in the next section.

3.2.7   Sensitivity Analysis

Sensitivity Analysis allows assessing a final ranking's robustness. A ranking is said to be robust if small variations of Criteria's weights do not cause changes in the ranking. This analysis is to be applied to all criteria but in the following, considering the example and without any loss of generality, the analysis will focus on the weight of the most important criteria, C_1 – "Interview result".

Firstly, the ranking is recreated in two parallel but extreme scenarios: what happens if the criterion being analyzed has its weight set to the maximum (scenario 1) and set to the minimum (scenario 2). The first scenario is very straightforward. If $w_1 = 1$ then all Alternatives' weighted sums will be equal to their corresponding scores in C_1 as can be verified in Table 3.7.

Scenario 2 involves an additional assumption. If $w_1 = 0$, then $w_1 + w_2 + w_3 < 1$ therefore the remaining weights have to be standardized, again. The analysis assumes the relation between the weights, besides the one being analyzed, is kept.

In the example: $w_1 + w_2 + w_3 = 0 + 0.4 + 0.1 = 0.5$, therefore the new standardized weights will be $w_2 = \dfrac{0.4}{0.5} = 0.8$ and $w_3 = \dfrac{0.1}{0.5} = 0.2$. The resulting sums are presented in Table 3.8.

Table 3.7: Sensitive Analysis – Scenario 1

	C_1	C_2	C_3	Sum
Weights	1	0	0	-
A	100	50	10	100
B	60	100	50	60
C	30	75	100	30
D	60	0	90	60
E	100	0	0	100

While comparing both Table 3.7 and Table 3.8 with Table 3.6 it is obvious that the ranking suffered some changes when the weight initially given by the DM, $w_1 = 0.5$, changed to the aforementioned scenarios. The issue here is to know for what values of w_1 can those changes in rank happen?

Table 3.8: Sensitive Analysis – Scenario 2

	C_1	C_2	C_3	Sum
Weights	0	0.8	0.2	-
A	100	50	10	42
B	60	100	50	90
C	30	75	100	80
D	60	0	90	18
E	100	0	0	0

Figure 3.9 shows the evolution of each Alternative's weighted sum in the range of $w_1 \in [0,1]$ and keeping the relation between w_2 and w_3 constant. As was expected, due to the dominance relation established in section 3.2.4 the line corresponding to Alternative E's sum is always below Alternative A's line.

The vertical dashed line represents the actual value of $w_1 = 0.5$ as chosen by the DM. As can be observed, if there is a small increase in w_1 there will be a rank inversion between Alternatives B and A, and between Alternatives E and C. To be exact, the mentioned rank inversions happen, respectively, for $w_1 = 0.55$ and $w_1 = 0.53$. That means the current ranking is not robust. Steps 5 and 7 should be revised until a robust ranking is obtained. Obviously, in the explored example, the second rank inversion is not that worrisome because only one person is going to be hired. But it is absolutely necessary to ensure that, for the 1st rank, the ranking is relatively insensitive to small fluctuations in any criterion weight.

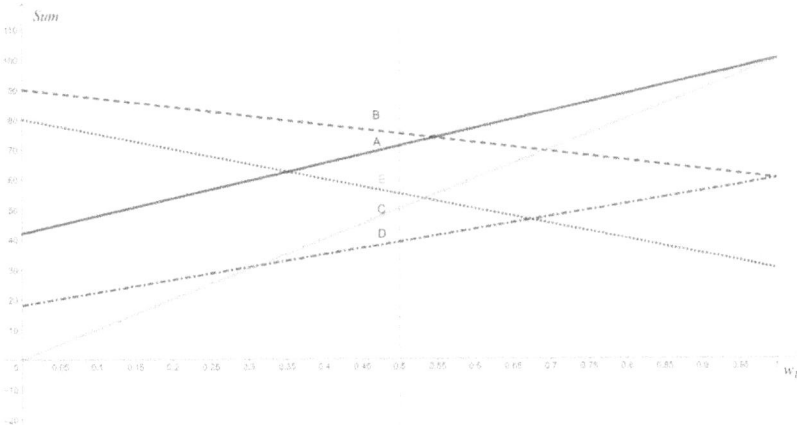

Figure 3.9: Sensitive Analysis to "Interview Result".

Note that, in this example, for $w_1 > 0.69$ Alternative E, dominated by A, will be in the 2nd rank. Again, dominated Alternatives should only be removed from the process if the circumstances allow it, for example when the problem is solely the selection of the best Alternative.

There are other techniques for Sensitivity Analysis, namely the TRIDENT analysis. This technique, proposed in Valadares et al. (1996) is limited to three criteria but gives more information concerning the relative relation of all criteria simultaneously. It makes it possible to visually evidence the set of weight's combinations for all possibilities of the best m Alternatives. Fig. 3.10 presents an example of a TRIDENT analysis with three Alternatives (A, B and C). The triangle represents the weight space and in each region is indicated the corresponding ranking. The final ranking, corresponding to the DM's actual set of weights is 1st – A, 2nd – B, 3rd – C. And that is a robust ranking, because small changes in any weight will not cause a rank inversion.

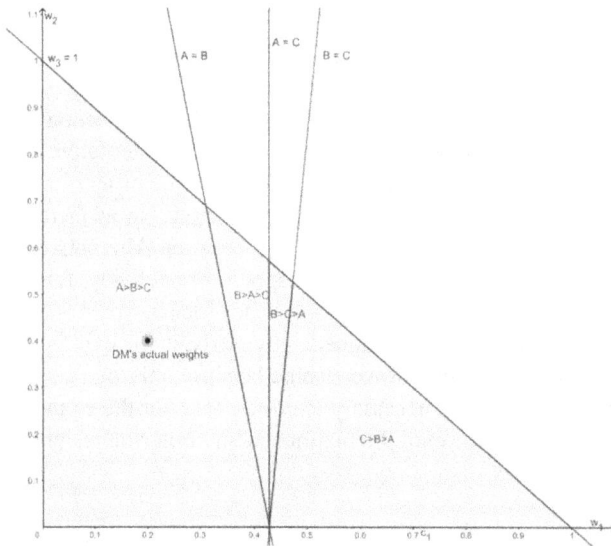

Figure 3.10: Example of a TRIDENT Analysis.

3.3   Final Remarks

This chapter does not describe the SMART technique as it was initially published by Edwards. It also includes the Swing Weight process that was introduced later with the SMARTS methodology. SMARTS also uses linear unidimensional utilities instead of Value Functions, making its procedures more easily understood by DMs and, according to Edwards, reducing possible assessing errors during step 4. Nevertheless, the presented version of SMART is still used for MCD problems due its simplicity. Although it heavily depends on the interaction with the DM for the Value Trees, the Value Functions and the Swing Weights. Any of the mentioned techniques, in spite of being simple to apply, can be hard to explain to a lay person and sometimes the results do not accurately reflect the Decision Maker's opinions and preferences simply because the procedures were not totally understood by the DM. And that is why the concept of robustness is so crucial in the final stage of SMART.

To emolliate SMART's, and in part SMARTS', limitations Edwards and Barron (Edwards & Barron, 1994) proposed SMARTER (SMART Exploiting Ranks). SMARTER also uses the linear unidimensional utilities used in SMARTS but implements the Swing Weight a bit differently. The DM is also asked to evaluate the various swings, like in SMARTS, but instead of directly assigning a percentage to each swing, the DM just has to rank them from the most important to the least. The procedure then uses the Rank Order Centroid (ROC) to transform those ranks into actual weights. That makes the DM's task much easier, when compared to SMARTS' Swing Weight. Nevertheless, Goodwin and Wright (Goodwin and Wright 2014) point out to some additional drawbacks these simplifications may cause and recommend that SMART(S) should not be considered an obsolete methodology.

3.4 Proposed Exercises

1. Gabriela wants to buy a new car and does not know where to begin. She is a very good friend of yours and you are willing to help her. The first question you asked her was: "What points of view do you consider important in a car?" After some time thinking, Gabriela told you that:

"For me, the most important aspect is the look! I want people to watch and say: 'What a nice pair of wheels she is driving'. Secondly, the price, the lower, the merrier. More money for fuel. Finally, the car's performance. The price is a bit less important than the looks, but the performance, although I found it relevant, is much less important than the other two issues."

Using a magazine you collected the information and organized it in Table 3.9.

Table 3.9: Information Concerning Six car Alternatives Evaluated in Criteria Price, Looks and Performance

Car	Price (in Plins)	Looks	Performance
A	350	Very Good	12
B	750	Very Good	18
C	750	Perfect	20
D	120	Awful	15
E	250	Good	12
F	350	Good	11

Unfortunately, Gabriela has a budget of 1,000 Plins, consequently some cars had to be removed from Gabriela's initial selection.

The cars' looks were attributed by Gabriela using the qualitative scale {Awful, Bad, Meh, Good, Very Good, Perfect}. After some questions that made her head spin you managed to learn that, for Gabriela, the increase of Value when changing from a car with a Bad look to a Meh look and from a Very Good look to a Perfect one is the double of the value difference between all other pairs of contiguous categories.

The cars' performances were taken from the magazine's evaluations on a scale from 0 (the worst) to 20 (the best).

Which car would you recommend to your friend?

2. Brutus, the brave and bold warrior earned a couple of silver pieces on his last adventure. He successfully saved the (not so) beautiful Princess Blimunda from the dreadful Dark Lagoon Monster. Unfortunately, the creature's acid destroyed Brutus' keen blade, Valeria. Back to the city, Brutus went to his friend Ferreira, the blacksmith, to buy a new weapon. Ferreira has some magic weapons available which have been evaluated by Brutus in three characteristics: Price, Damage and Ease of use. All information has been gathered in Table 3.10.

Brutus does not intend to buy a weapon that does less than 5 hit points. He has a very linear mind and considers all the mentioned attributes equally important.

Help Brutus pick his new weapon of choice. Perform a Sensitive Analysis to evaluate your suggestion's robustness.

3. Evarista, the Witch, has a situation. She has to run away from her coven's town because the other witches discovered she is doing good deeds for the town folks. When she applied for the place she had no idea it was an Evil Coven! And now the other crones are out there to get her. Fortunately, she has been doing this witch's life for decades and she is prepared for these kinds of circumstances. She has several caches in different towns so she can start again somewhere else. The problem is, escaping is risky and she does not know what town she prefers to

Table 3.10: Evaluation of Ferreira's Available Weapons in Criteria Price, Damage and Easiness of Use

Weapon	Price (in silver pieces)	Damage (in hit points)	Ease of use
Alice (dagger)	2	4	Extremely Easy
Butch (mace)	5	6	Very Easy
Crunch (morningstar)	8	8	Easy
Dalila (spear)	6	6	Easy
Erving (sword)	15	8	Hard
Fandango (axe)	20	12	Very Hard

escape to. After she gathered her thoughts she settled the following 6 towns that were assessed according to Cache saved, probability of getting caught and town's pleasantness (Table 3.11).

Table 3.11: Information Collected by Evarista Concerning Possible Escape Towns

Town	Cache (in Gold Pieces)	Probability of Getting Caught	Town's Pleasantness
Altamira	700	0.05	High
Bayona	1500	0.25	Low
Cantlecaster	1300	0.25	Null
Doh	240	0.00	Average
Eboricum	1600	0.55	High
Ferim	500	0.10	High

Evarista does not want to take too high a risk. She will not try to escape to a town with a probability of being caught greater than 30%. The pleasantness Attribute was measured in the qualitative scale (Null, Very Low, Low, Average, High, Very High).

(a) Recommend a destiny to Evarista considering linear Value Functions and equal weights for the three criteria.

(b) Implement Sensitive Analyses to assess your suggestion's robustness.

4. Auntie Leonarda needs to hire a private jet plane for crossing the continent and go to an important event, tomorrow. She checked a site on-line and pre-selected the four models that pleased her most. Table 3.12 presents all information Auntie Leonarda considers relevant for her final decision.

The Attribute "Comfort" was assessed by previous customers using the qualitative scale: Terrible; Very Bad; Bad; Average; Good, Very Good, Excellent.

(a) Consider D the increase in value, from Auntie Leonarda's perspective, when she can change from a plane with a terrible comfort level to one with a very bad comfort. After a conversation with Auntie Leonarda it was settled that the increase in value in a change from a plane with Very Good Comfort to one with Excellent comfort is $4D$, the same increase when changing from Good to Very Good is $3D$, and going from Average to Good is $2D$. The remaining value increases are approximately D. Build an adequate Value Function that incorporates Auntie Leonarda's preferences.

(b) Assume linear Value Functions for the other Attributes. Rank the four models knowing that Auntie Leonarda considers "Comfort" the most important criterion, immediately followed by "Trip Duration". "Cost" is considerably less important than the other two.

5. The Government of Lusoland is considering the viability of building a third bridge across river Tajus in Lisbonia metropolitan area. There are four possible

Table 3.12: Information gathered by Auntie Leonarda

Car Model	Cost (in Monetary Units) C_1	Trip Duration (in Hours) C_2	Comfort C_3
Amethyst (A)	3000	6	Very Good
Beryl (B)	1500	6 h 30 min	Average
Crystal (C)	1000	10	Good
Diamond (D)	5000	4 h 15 min	Excellent

places where the bridge can be built. The four alternatives were evaluated according to three points of view: Cost (in millions of Eurons); Environmental Impact (in a scale of 0 - null to 20 - catastrophic); and Road Traffic relief (in a qualitative scale "Null, Very weak, Weak, Average, High, Very High, Extreme").

The information concerning the evaluations was collected and is presented in Table 3.13.

Table 3.13: Information Gathered Concerning Four Possible Crossing Locations.

Possible crossing	Cost	Environmental Impact	Traffic Relief
Arreiro – Lisbonia	300	9	Average
Brandão – Lisbonia	150	6	Average
Ceixal – Lisbonia	100	5	Weak
Drafaria – Lisbonia	500	18	Extreme

(a) Consider that the value increase corresponding to a change from an Average traffic effect to an Extreme one is 80% of the value increase of changing from a weak effect to an extreme effect. Built a Value Function that verifies this condition and calculates the Values of each Alternative in the criterion "Traffic Effect".

(b) Assume linear Value Functions for the other two criteria. Rank the four Alternatives assuming all criteria have the same weight.

(c) For political reasons, the Government intends to build the bridge connecting Drafaria to Lisbonia. Find a set of weights that ensure this Alternative is in the 1st rank and evaluate the obtained ranking's robustness.

6. Auntie Leonarda wants to throw a tremendous party to celebrate the beginning of the Social Season. Zumélia, her niece and assistant, suggested some possible themes for the party: the Aqua Party (a traditional celebration by the artificial lake in Auntie's mansion grounds, lots of fountains and waterfalls); the Black Party (everyone dressed in elegant black clothes, dark decoration and black food and beverages); the China Party (Far East themed decoration, Oriental inspired music and clothes, Chinese food); the Divine Party (guests must dress like Greek or Egyptian Mythology characters, ethereal and grandiose decoration).

"To help you out I organized all information in Table 3.14. It considers my assessment of the four alternatives from three criteria's perspective: party cost (in monetary units), impact in the jet set society (the number of celebrities who probably will attend the party), and organization difficulty (from 'Very Easy' to 'Very Complex')" – mentioned Zumélia.

Table 3.14: Information Organized by Zumélia

Party	Cost	Jet Set Impact	Difficulty
Aqua	2,000	60	Complex
Black	1,200	35	Very Easy
Chinese	500	40	Easy
Divine	3,000	120	Average

Auntie Leonarda considers that the value increase from changing from an "average" to organize party to a "complex" party should be four times bigger than the value difference between an "average" and an "easy" one.

(a) Assuming equal weights for the three criteria, obtain a ranking for the four parties.
(b) Implement a sensitivity analysis to each criterion and comment on the robustness of the ranking obtained in a).

7. The Mayor of Vila Nova de Carambola intends to build a new landfill in the municipality. Obviously, whatever place is chosen will cause protest and loss of popular support. So, the decision has to be carefully taken. There are four available sites that were already evaluated from three points of view: Distance to the main town (the further, the better, out of sight, out of mind); installation cost (in monetary units) and environmental impact that was measured in a qualitative scale ("Very High", "High", "Average", "Low", "Neglectable").

The evaluations' results can be consulted on Table 3.15.

Table 3.15: Evaluation of Four Possible Sites to Build the
Vila Nova de Carambola New Landfill

Site	Distance (in km)	Cost	Environmental Impact
W	25	300	High
X	12	125	Low
Y	5	200	Low
Z	30	600	Very High

The Mayor considers that the value difference between two consecutive categories from the Environmental Impact qualitative scale, should be the double of the difference between the immediately lower categories. For example, the difference of value between "Very High" and "High" should be the double of the difference between "Average" and "High". The Mayor also thinks that the Attribute Cost is the most important. The Attribute Distance is just slightly less important than Cost. Attribute Environmental Impact is frankly less important than Cost.

Help the Mayor pick the new location. Perform a Sensitive Analysis to evaluate your suggestion's robustness.

8. Auntie Leonarda is planning her summer vacations and although she has not decided where to go, she knows she will be taking a nephew or niece with her. She wants to take Alzira, but she must "scientifically" justify her decision or she will have a jealousy riot amongst the gang. Therefore, she decided to consider the following aspects:

- The nephew/niece' last semester grades (the higher, the better);
- The number of trips each youth has already done with her (the lower, the better, Balancing karma, my dears);
- The value of the gift each of them gave Auntie on her last Birthday (the higher the better, Time for paying back).

Table 3.16 presents the information collected by Auntie.

Table 3.16: Information Collected by Auntie Leonarda

Nephew/Niece	Grades Average (0 – 20)	# of previous trips	Gift value (in plins)
Alzira	15.6	3	10,500
Bernardo	16.2	5	9,000
Cecília	14.3	2	6,500
Duarte	16.0	5	7,500
Ermelinda	15.1	3	10,000

(a) Assume linear Value Functions and equal weights for the three criteria. Obtain the final ranking.
(b) Propose weights and Value Functions that ensure Alzira will be in the 1st rank of the final ranking.
(c) Compare the robustness of rankings obtained in (a) and (b)

9. Jill and Jane (J&J) are buying a house. After a long market consultation, they selected the most interesting options. All have similar areas and building quality. The choice will be made based on:

- Price, in plins;
- Average distance from their workplaces, in km;
- Pleasantness, evaluated through Direct Rating using a 0–20 scale.

J&J organized all relevant information in Table 3.17.

J&J consider the Distance to their works the most important criterion. Price and Pleasantness are equally important, but less than the Distance criterion.

Table 3.17: Information Collected by J&J

Alternative	Price	Distance	Pleasantness
W	3,000	30	16
X	5,000	10	14
Y	9,000	15	20
Z	9,000	50	13

Assuming linear Value Functions, obtain the final ranking. Analyze the ranking's robustness.

References

Edwards, W. (1971). Social utilities. *The Engineering Economist*, 6, 119–129.

Edwards, W. and F.H. Barron (1994). Smarts and smarter: Improved simple methods for multiattribute utility measurement. *Organizational Behavior and Human Decision Processes*, 60(3), 306–325. https://doi.org/10.1006/obhd.1994.1087

Goodwin, P. and G. Wright (2014). *Decision Analysis for Management Judgment*, 5th Edition. New York: John Wiley & Sons Inc.

Valadares Tavares, L., F. Nunes Correia, I.H. Themido and R.C. Oliveira (1996). *Investigação Operacional*. McGraw-Hill.

von Winterfeldt, D. and W. Edwards (1986). *Decision Analysis and Behavioral Research*. Cambridge University Press.

The ELECTRE Methods

4.1   Introduction

The ELECTRE (*ELimination Et Choix Traduisant la REalité* – Elimination and choice translating the Reality) is a MCD methodology created by Bernard Roy in 1968 (Roy 1968). Later, this methodology was expanded by Roy and other authors and several variants were developed such as ELECTRE II, ELECTRE III, ELECTRE IV, ELECTRE IS, ELECTRE TRI, among others.

It is a non-compensatory approach to MCD problems and it tries to compensate for one of the drawbacks of compensatory additive methods, like SMART and its variants. In a compensatory method, given a set of Alternatives ranked according to a set of Criteria, if one Alternative has a very bad score in one criterion, this could be compensated by an excellent score in a different criterion. And, depending on the criteria's weights, that particular Alternative could end up in the top rank. Non-compensatory approaches try to prevent these situations from happening, imposing a veto when the result in one criterion is below an acceptable level.

ELECTRE methodology focuses not on the Alternatives' scores in the different criteria, but on their scores' differences. These differences are incorporated in matrices that will be used to evaluate the quality of the relation between each pair of Alternatives. This method prevents that badly scored Alternatives in, at least, one criterion may raise to the top ranks, as was mentioned above. But, because it deals with the scores differences, the method does not have to worry about rescaling the scores and can handle qualitative scales without transforming them into numerical values.

This chapter will explore deeper the ELECTRE III version. Nevertheless, the Section 4.2 will address ELECTRE I because the first variant introduced core concepts that were later used by all ELECTRE methods. The definition of Outranking Relation will be presented in Subsection 4.2.1. Afterwards, the two antagonist forces that influence the final rankings in ELECTRE are explored: the Concordance (Subsection 4.2.2) versus Discordance and Veto (Subsection 4.2.3). Finally, the Nucleus, the most important output of this method, is presented in Subsection 4.2.4.

Section 4.3 has the bulk of the chapter. It presents ELECTRE III in detail. The Preference and Indifference are introduced in 4.3.1 and afterwards the Outranking Relation is revisited and given an ELECTRE III framing (4.3.2). Subsections 4.3.3 and 4.3.4 explore the Concordance and Discordance Matrices using the Preference and Indifference Relations mentioned above followed by the presentation of the Credibility Matrix in Subsection 4.3.5. Finally, in Subsection 4.3.6, a method that can be used to extract the Alternatives' final ranking is detailed.

The chapter concludes with some final remarks concerning some of the method's drawbacks (section 4.4) and a collection of recommended exercises (Section 4.5).

4.2 ELECTRE I

This method's purpose is to extract, from a set of Alternatives, the smallest subset of the most promising compromise of Alternatives. This subset is called the Nucleus. It is a particularly useful method when there are a high number of Alternatives and therefore can be used in tandem with a different methodology that is later applied to the Nucleus. ELECTRE I was not designed to rank all alternatives in the Nucleus, and that fact led to the creation of the later and most complex ELECTRE variants.

In ELECTRE I the Alternatives' quality is evaluated by pairwise comparisons. These comparisons consider two possibly antagonist perspectives: the Concordance and the Discordance over an outranking relation between the two Alternatives being compared.

4.2.1 Outranking Relation

Given two Alternatives A and B, A is said to outrank (or to subordinate) B if A is, at least, as good as B in all criteria. From a different perspective A outranks (or subordinates) B if it is not worse than B in all criteria. It is represented as ASB, with S meaning Subordination.

For any pair of alternatives A and B four possible situations may arise:

- A outranks B, but B does not outrank A, *i.e.* ASB and $\sim BSA$;
- B outranks A, but A does not outrank B, *i.e.* BSA and $\sim ASB$;
- A outranks B and B outranks A, *i.e.* ASB and BSA;
- A does not outrank B and B does not outrank A, *i.e.* $\sim ASB$ and $\sim BSA$.

In the first two situations there is an obvious Alternative that outperforms the other. In the third situation both Alternatives outrank each other. This represents a situation when there is indifference between the two Alternatives and they are said to be indifferent. The last situation addresses the occasion when neither Alternative outranks the other. This indicates that there is an incomparability and the two Alternatives are said to be incomparable.

To assess the outranking relation between two Alternatives, ASB, ELECTRE considers two principia:

- The Concordance Principium, that demands that A is not worse than B for a majority of criteria;
- The Non Discordance Principium, which requires that in the minority of criteria that do not support the assessed outranking relation there is not a strong rejection of that relation. Meaning, A can be worse than B in a certain criterion but can not be too much worse.

The issue here is how to quantify what is considered "too much worse"?

The outranking relation is quantified using the aforementioned principia and considering weights that were previously assigned to the criteria. These weights do represent an absolute degree of importance for the DM and not a relative one as happens for the Compensatory methodologies. The weights do not have to add to 1 when they are being assigned by the DM, but later in the process it will be necessary to standardize them.

Consider the example presented in the Appendix B:

Table 4.1: Example Data

		Interview	Demerits in Practical Test	Theoretical Test
	A	Very Good	4	75.4%
	B	Good	2	84.6%
Candidates	C	Regular	3	95.7%
	D	Good	6	90.2%

Consider that for the DM the most important Criterion, followed by the number of Demerits in the Practical test. The DM assigned the weights $w_1 = 3$, $w_2 = 2$ and $w_3 = 1$. Standardizing the weights will get:

$$w_1 + w_2 + w_3 = 6 \Rightarrow w_1 = \frac{3}{6} = 0.5, \quad w_2 = \frac{2}{6} = 0.33, \quad w_3 = \frac{1}{6} = 0.17$$

Table 4.2: Criteria's Weights

Interview	Demerits in Practical Test	Theoretical Test
0.5	0.33	0.17

4.2.2   Concordance Matrix

In order to assess the outranking relations between each pair of Alternatives it is necessary to firstly assess the concordances, meaning to assess how much the criteria supports each outranking relation. Those concordance values are recorded in the Concordance Matrix. The element a_{ij} of the Concordance Matrix is also represented by $C(i, j)$ and it quantifies how much the criteria supports the relation iSj where (i, j) is a pair of Alternatives. The value a_{ij} is equal to the sum of the weights from the criteria that support the relation iSj.

In the example, for *ASB*, Alternative *A* has a better result in the interview than Alternative *B*. So this criterion is agreeing with the relation *ASB*. On the other hand, candidate *A* had more demerits in the practical test and has a lower grade in the written test than the candidate *B*, therefore the later criteria are not in concordance with the relation *ASB* and $C(A, B) = 0.5 + 0 + 0 = 0.5$.

Symmetrically, for $C(B, A) = 0 + 0.33 + 0.17 = 0.5$.

It might seem that symmetrical positions in the matrix will add up to 1, but due to the way the outranking relation is defined, that is not always the case.

For example, for *BSD* and *DSB* both candidates had the same result in the interview, candidate *B* had less demerits than *D* but *D* had a better grade in the written test. Therefore:

$$C(B, D) = 0.5 + 0.33 + 0 = 0.83$$
$$C(B, D) = 0.5 + 0 + 0.17 = 0.67$$

Obviously, $C(B, D) + C(D, B) = 0.83 + 0.67 > 1$. This situation occurs when two Alternatives have the same score in a given criterion.

The complete Concordance Matrix for the example is presented in Table 4.3.

Table 4.3: Concordance Matrix

	A	B	C	D
A	1	0.50	0.50	0.83
B	0.50	1	0.83	0.83
C	0.50	0.17	1	0.50
D	0.17	0.67	0.50	1

4.2.3 Discordance and Veto

The Concordance Matrix corresponds to the first Principium, and the Discordance will address the second. And for the Discordance analysis there is a fundamental notion: the Veto Condition. The Veto Condition is triggered when, for a given criterion, the Alternative *i* is frankly worse than Alternative *j*. In that case, not only is that criterion is not in concordance with *iSj* but it might trigger a Veto. A Veto will prevent that $C(i, j) > 0$ even if *i* is better than *j* in the other criteria. In some circumstances the relation *iSj* can be vetoed by more than one criteria, but it does not matter. Even if the relation *iSj* is only vetoed by one criterion then $C(i, j)$ will be zero. The issue here is how to assess if one Alternative is frankly worse than the other? For that the DM is asked to define Veto Thresholds for all criteria. Each Veto Thresholds is defined by the DM taking into consideration the differences between the Alternatives' scores on the corresponding criterion.

In the example, the criterion Interview was assessed through a qualitative scale with six categories. The veto condition could be triggered for *iSj* if *j* is better than *i* by at least two categories. For the practical test, assessed by the number of obtained demerits, the veto threshold could be 3, meaning that the second criterion vetoes *iSj* if candidate *i* had 3 or more demerits than candidate *j*. Table 4.4 presents the Veto Thresholds for the three criteria.

Table 4.4: Criteria's Veto Thresholds

Interview	Demerits in Practical Test	Theoretical Test
3 categories	3	15%

After the veto thresholds are settled it is possible to obtain the Veto Matrix. For every possible outranking relation the method will check if any Veto Condition is triggered and in that case will register it in the matrix.

For example, considering alternatives *B* and *D*, the relation *BSD* has a Concordance of 0.83 because the only criterion than is not supporting the relation is the third, where *D* is better than *B*. But the score difference is just $90.2 - 84.6 = 5.6\% < 15\%$ therefore this criterion will not veto the relation. On the other hand, for *DSB*, the criterion not supporting the relation, the second, the difference is $6 - 2 = 4$ that is greater or equal than the veto threshold for this criterion, therefore criterion 2 will veto the relation *DSB*.

Table 4.5 presents all vetoes found in the example.

Table 4.5: Vetoes Matrix

	A	B	C	D
A	-	No Vetoes	Veto: 3rd Crit.	No Vetoes
B	No Vetoes	-	No Vetoes	No Vetoes
C	No Vetoes	No Vetoes	-	No Vetoes
D	No Vetoes	Veto: 2nd Crit.	Veto: 2nd Crit.	-

The issue now is how to integrate the two matrices into one, meaning: how to relate the Concordance Matrix with the Veto Matrix? Firstly, the vetoes should be applied and the corresponding Concordance values should be considered to be null. In Table 4.6 the concordances of the vetoed relations are stroked. For example, in spite of its considerable high concordance value of 0.67, relation *DSB* was vetoed by criterion "Demerits in practical test", therefore the value is considered zero and the outranking relation is rejected, ~*DSB*.

Table 4.6: Vetoes Affecting the Concordance Matrix

$$C(i, j) = \begin{array}{c} \\ A \\ B \\ C \\ D \end{array} \begin{array}{cccc} A & B & C & D \\ \begin{bmatrix} 1 & 0.50 & 0.50 & 0.83 \\ 0.50 & 1 & 0.83 & 0.83 \\ 0.50 & 0.17 & 1 & 0.50 \\ 0.17 & 0.67 & 0.50 & 1 \end{bmatrix} \end{array}$$

What can be said about the remaining outranking relations? $C(A, D) = 0.83$, but $C(D, A)$ is just 0.17. Although it seems obvious that ASD, what happens to the relation DSA?

An outranking relation is only be accepted if its concordance, after veto, is greater or equal to a predetermined value: the Concordance Threshold, also denominated the Outranking Threshold, and represented by $\hat{C}$.

Finally it is possible to formally define the outranking relation.

Given two Alternatives i and j, it is said that i outranks $j(iSj)$ if and only if $C(i, j) > \hat{C}$ and no criteria is vetoing the relation. Or:

$$iSj \Leftrightarrow C(i, j)_{after\ veto} \geq \hat{C}$$

For the example, and considering $\hat{C} = 0.6$, we can establish the relations presented in Table 4.7.

Table 4.7: Outranking Matrix

$$\begin{array}{c c c c c}
 & A & B & C & D \\
A & - & \sim ASB & \sim ASC & \sim ASD \\
B & \sim BSA & - & BSC & BSD \\
C & \sim CSA & \sim CSB & - & \sim CSD \\
D & \sim DSA & \sim DSB & \sim DSC & 1
\end{array}$$

The only outranking relations of the problem is: ASD, BSC and BSD. And now?

4.2.4 The Nucleus

The last step of ELECTRE I is finding the Nucleus of the Alternatives set. The Nucleus is the smallest subset of Alternatives where any Alternative in the Nucleus is not outranked by any Alternative that does not belong to the Nucleus. In other words, every Alternative that does not belong to the Nucleus is outranked by, at least, one Alternative in the Nucleus.

As was mentioned in the previous section, the only outranking relations in the example are: ASD, BSC and BSD. So, what is the Nucleus of $X = \{A, B, C, D\}$?

Well, X always verifies the outranking condition, because there are no Alternatives "outside" of X. But it is not an interesting answer because the Nucleus is the subset with the lowest cardinality that verifies the condition.

For example, X admits four subsets of cardinality 3: $\{A, B, C\}$; $\{A, B, D\}$; $\{A, C, D\}$ and $\{B, C, D\}$. But only two of them verify the outranking condition. $\{A, C, D\}$ can not be the Nucleus because $B \notin \{A, C, D\} \land BSD$

Similarly, $\{B, C, D\}$ cannot be the Nucleus because $A \notin \{B, C, D\} \land ASD$.

Notice that both $\{A, B, C\}$ and $\{A, B, D\}$ verify the outranking condition. For example, $D \notin \{A, B, C\} \land \sim DSA, \sim DSB, \sim DSC$.

Below is presented the list of all possible subsets of X and which of them verify the outranking condition:

$$X; \{A, B, C\};$$

In conclusion, the example admits that two different subsets can be Nucleus: $\{A\}$ and $\{B\}$. And because these Alternatives are incomparable, this is not the ideal situation. Sometimes, changing the Threshold values by a little will help the identification of a single Nucleus.

Usually, representing the outranking relations in a graph will tremendously help in finding the Nucleus. Figure 4.1 presents the example's outranking graph, where it is possible to visualize the two possible nucleuses with cardinality 1. Figure 4.2 illustrates an example with the ideal situation, with a single possible Nucleus with a cardinality 1.

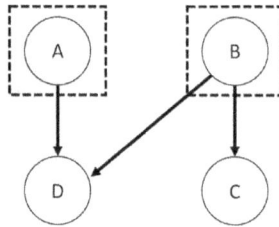

Figure 4.1: Example's Outranking Graph

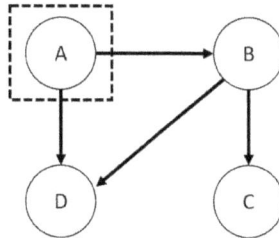

Figure 4.2: Outranking Graph – Single Nucleus

In other situations, the Nucleus has a cardinality higher than 2. And, in this case it will not be possible, using solely this method, to recommend an Alternative to the DM, because ELECTRE I will be able to rank the Alternatives in the Nucleus. Figure 4.3 exemplifies such a situation.

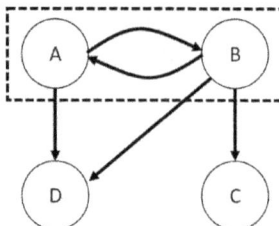

Figure 4.3: Outranking Graph - Nucleus with Cardinality 2

These are ELECTRE I's major drawbacks and they were corrected in the later variants as will be shown in the next section.

4.3   Electre III

ELECTRE III is a more complex and sophisticated variant of the original ELECTRE presented in section 4.2. Roy created it (Roy 1978) to overcome the most obvious ELECTRE I's drawback: the inability to rank the most interesting alternatives. This variant introduces other thresholds and the notions of Weak and Strong preference.

4.3.1   Preference and Indifference Relations

Given a and b, two generic Alternatives in a Multicriteria Decision Problem, and a Criterion with a scoring function g it is natural to define the following relations:

- a is Preferable to b (aPb) if $g(a) > g(b)$ and
- a and b are Indifferent (aIb) if $g(a) = g(b)$.

However, the way these relations are defined can be highly unrealistic. What happens if the scores of a and b are very similar with one being only slightly greater than the other? Are there really a Preference relation between them or is it possible to consider them being Indifferent?

To answer these questions ELECTRE III introduces the concept of Indifference Threshold, represented by q that will allow to define more realistically the relations presented above.

Given a and b, two generic Alternatives in a Multicriteria Decision Problem, a Criterion with a scoring function g and an Indifference Threshold of q:

- a and b are Indifferent (aIb) if $|g(a) - g(b)| \leq q$

In other words, two Alternatives are considered Indifferent relative to a Criterion if the difference of their scores in that Criterion is lesser or equal than the Criterion's Indifference Threshold.

The Indifference Relation is obviously symmetrical, $aIb \Rightarrow bIa$.

The Indifference Thresholds are defined by the DM for all Criteria and depend on the scale being used for each Criterion and, therefore, it is natural for a MCD problem to have very diverse Indifference Thresholds.

A similar reasoning allows ELECTRE III to evaluate if the difference between two Alternatives' scores is "very high" or "just a bit high" and these nuanced qualifications lead to a refinement to the concept of Outranking already explored in ELECTRE I. The Preference Thresholds (p) are also defined by the DM for each Criterion and will allow the assessment of preference relations' intensity.

Given a and b, two generic Alternatives in a Multicriteria Decision Problem, a Criterion with a non-decreasing scoring function g, an Indifference Threshold of q and a Preference Threshold of p:

- a is Strongly Preferable to b (aPb) if $g(a) - g(b) > p$ and
- a is Weakly Preferable to b (aQb) if $q < g(a) - g(b) \leq p$.

In other words, there is a Strong Preference between two Alternatives if the difference between their scores is so high that is greater than the Preference Threshold. If the difference is neither "too small" nor "too large", meaning it is between the Indifference Threshold and the Preference Threshold, then there is a Weak Preference between the Alternatives.

Obviously that, for every Criterion, $p > q$.

In summary

Given a and b, two generic Alternatives in a Multicriteria Decision Problem, a Criterion C_j with a non-decreasing scoring function g_j, an Indifference Threshold of q_j and a Preference Threshold of p_j it is possible to define the following relations:

- a is Strongly Preferable to b (aP_jb) if $g_j(a) - g_j(b) > p_j$;
- a is Weakly Preferable to b $(aQ_j b)$ if $q_j < g_j(a) - g_j(b) \leq p_j$;
- a and b are Indifferent (aI_jb) if $|g_j(a) - g_j(b)| \leq q_j$.

Figure 4.4 shows the possible relations between two Alternatives, a and b, depending on their scores' difference, from a Criterion perspective, given its Indifference and Preference Thresholds, q_i and p_i.

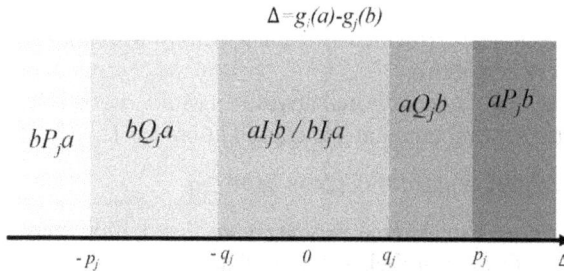

Figure 4.4: Indifference, Weak and Strong Preference Relations

Notice that these relations' definitions assume that the scoring function is non-decreasing. If a Criterion has a decreasing scoring function then the Strong and Weak Preference Relations' definitions would have to be adjusted. For simplicity, in that situation, the authors recommend that the Alternatives' scores should be multiplied by (–1) to invert the scale and that will make the above definitions valid.

Also, notice that although the relation P is transitive, Q and I may not be. It is very common to happen aIb, bIc and aQc. Similarly, aQb, bQc and aPc. It depends considerably from the Criterion's q and p.

As was mentioned before, for all MCD methods it is important to incorporate the DM's subjectivity so the analysis reflects the DM's preferences and biases. For

ELECTRE, the choice of the Thresholds is fundamental for a realistic modeling of that subjectivity and therefore should be done thoughtfully. For additional and deeper study concerning Thresholds definition in ELECTRE read (Rogers and Bruen 1998).

Example

Observing the example:

Table 4.8: Example with Indifference and Preference Thresholds

		C_1	C_2	C_3
		Interview	**Demerits in Practical Test**	**Theoretical Test**
	p_j	1 category	2 demerits	12%
	q_j	0 categories	1 demerits	6%
Candidates	A	Very Good	−4	75.4%
	B	Good	−2	84.6%
	C	Regular	−3	95.7%
	D	Good	−6	90.2%

For Criterion C_1, which was scored through a qualitative scale, the DM decided two Alternatives would only be Indifferent if they had the same score. And there will only be a Strong Preference if they have a difference of more than two categories. The scores in C_2 were multiplied by (−1) because the scoring function is decreasing. The higher the number of demerits the worse is the DM's opinion for that candidate.

Analyzing the scores' differences for each pair of Alternatives and for each Criteria:

For C_3 – Grade in the theoretical written test:

- *A, B* B has a better grade and $g_3 (B) - g_3 (A) = 84.6 - 75.4 = 9.2$, $6 < 9.2 \le 12 \Rightarrow BQ_3A$,
- *A,C* C has a better grade and $g_3 (C) - g_3 (A) = 95.7 - 75.4 = 20.3$, $20.3 > 12 \Rightarrow CP_3A$,
- *A, D* D has a better grade and $g_3 (D) - g_3 (A) = 90.2 - 75.4 = 14.8$, $14.8 > 12 \Rightarrow DP_3A$,
- *B, C* C has a better grade and $g_3 (C) - g_3 (B) = 95.7 - 84.6 = 20.3$, $6 < 9.2 \le 12 \Rightarrow CQ_3B$,
- *B, D* D has a better grade and $g_3 (D) - g_3 (B) = 90.2 - 84.6 = 5.6$, $5.6 \le 6 \Rightarrow DI_3B$, and $BI_3 D$,
- *C, D* C has a better grade and $g_3 (C) - g_3 (D) = 95.7 - 90.2 = 5.5$, $5.5 \le 6 \Rightarrow CI_3 D$, and DI_3C

For C_2: BQ_2A; AI_2C (and CI_2A); AQ_2D; BI_2C (and CI_2B); BP_2D; CP_2D

For C_i: AQ_1B; AP_1C; AQ_1D; BQ_1C; BI_1D (and DI_1B); DQ_1C

And now? What should be done with these relations? How to integrate them into a single measure of a "general preference" between two Alternatives? Let's recall some concepts introduced in the previous section.

4.3.2   Outranking Relation

As was mentioned for ELECTRE, given a Criterion C_j, aS_jb (a outranks b relatively to C_j) if a is not worse than b relatively to that Criterion. Meaning that C_j is in concordance with the relation aSb (a outranks b in general) if and only if aS_jb, which means that aP_jb, or aQ_jb or aI_jb. Meaning that even if $g_j(b)$ is slightly better than $g_j(a)$, if $g(b) - g(a) \leq q_j$, then aS_jb.

In summary, the following propositions are all equivalent:

- C_j is in concordance with the relation aSb
- aS_jb
- $aP_jb \; \dot\vee \; aQ_jb \; \dot\vee \; aI_jb$
- $g_j(b) - g_j(a) \leq q$

On the other hand, $\sim aS_jb$ means that b is preferable to a, or, in other words: $g_j(b) - g_j(a) > q_j$.

Another important definition: C_j is said to be in discordance with aSb if and only if bP_ja. Which is equivalent to $g_j(b) - g_j(a) > p_j$. Meaning that b has to be much better than a relatively to C_j for the Criterion to be in discordance with the Outranking relation aSb.

In summary, the following propositions are as equivalent:

- C_j is in discordance with the relation aSb
- $bP_j a$
- $g_j(b) - g_j(a) > p_j$

Notice that $\sim aS_jb$ does not necessarily imply that C_j is in discordance with aSb!

If bQ_ja, than $\sim aS_jb$, but $q_j < g_j(b) - g_j(a) \leq p_j$, meaning that in this particular situation, the Criterion C_j in neither in concordance nor in discordance with aSb.

Example

Consider the relation CSD. In Section 4.3.1 it was seen that DQ_1C, CP_2D and CI_3D. Therefore both C_2 and C_3 are in concordance with CSD. But, although C_1 is not in concordance with the relation, the Criterion is also not in discordance in spite of $g_1(D) > g_1(C)$, because DQ_1C.

But, relatively to the relation DSC, C_1 and C_3 are both in concordance, and C_2 is in discordance.

Knowing which Criteria are in concordance and in discordance with each outranking relation will allow the aggregation of all these relations in two different kinds of matrices: the Concordance Matrix, and the Discordance Matrices. The first will include the Principium of Concordance in the analysis. The latter will introduce the Principium of Non Discordance's perspective.

4.3.3 Concordance Matrix

Now, the method will evaluate the strength of all Outranking Relations and present the results in a Matrix. The element a_{ij} of the Concordance Matrix aggregates the contribution of each Criterion depending on if it is in Concordance with the Outranking Relation *iSj* or not. If a Criterion is in discordance with *iSj* then its contribution will be zero.

But before that aggregation is possible, it is necessary to assign an Importance Coefficient to each Criterion, according to the DM's preferences[1].

Let's denominate k_j as the Importance Coefficient of Criterion C_j. The Concordance level of the Outranking Relation *aSb* is represented by $C(a,b)$ and is defined by:

$$C(a, b) = \frac{\sum_{j=1}^{r} k_j c_j(a, b)}{K}$$

where r is the number of Criteria, $K = \sum_{j=1}^{r} k_j$, and $c_j(a, b)$ is the contribution of each Criterion's concordance to the relation *aSb*. The value of $c_j(a, b)$ is:

$$C_j(a, b) = \begin{cases} 1 & \text{if } g_j(b) - g_j(a) \le q_j \\ 0 & \text{if } g_j(b) - g_j(a) > p_j \\ \dfrac{p_j - (g_j(b) - g_j(a))}{p_j - q_j} & \text{othewise} \end{cases}$$

Notice that $C_j(a, b)$, the contribution of Criterion C_j for the concordance level of the Outranking Relation *aSb*, can be:

- 1, if and only if C_j is in concordance with the relation, i.e., if ap_jb, or aQ_jb, or aI_jb (and bI_ja);
- 0, if and only if C_j is in discordance with the relation, i.e., if bP_ja;
- A value in [0,1], if C_j is neither in concordance nor in discordance, i.e., if bQ_ja.

Figure 4.5 presents the graphic of $C_j(a, b)$. Notice that when bQ_ja, $C_j(a, b) = \dfrac{p_j - (g_j(b) - g_j(a))}{p_j - q_j}$ corresponds to a linear interpolation between the points $(q_j, 1)$, $(p_j, 0)$.

---

[1] As was mentioned for ELECTRE I, the Importance Coefficients are a concept that, although seeming similar, are extremely different from SMART Criteria's weights. In SMART, the weights assess the relative importance between the criteria and correspond to trade-off rates between them. In ELECTRE III this compensatory nature between the criteria does not exist. The Importance Coefficients solely represent the absolute importance the DM assigns to each criterion.

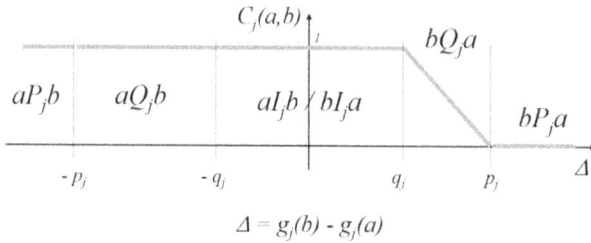

$$\Delta = g_j(b) - g_j(a)$$

Figure 4.5: Graphic Representation of Image $C_j(a, b)$

In summary, the value of $C(a, b)$ is the weighted sum of $C_j(a, b)$ using the Importance Coefficients as weights.

Example

Assume that the DM assigned 5, 3 and 2, respectively, to the Importance Coefficients of Criteria C_1, C_2 and C_3 as can be observed in Table 4.9.

Table 4.9: Example with Importance Coefficients

Interview		C_1	C_2	C_3
			Demerits in Practical Test	Theoretical Test
		5	3	2
	p_j	1 category	2 demerits	12%
	q_j	0 categories	1 demerits	6%
Candidates	A	Very Good	−4	75.4%
	B	Good	−2	84.6%
	C	Regular	−3	95.7%
	D	Good	−6	90.2%

Let's evaluate $C(C, D)$ and $C(D, C)$.

Previously, in section 4.3.2, it was settled that both C_2 and C_3 were in concordance with *CDS*. But there were no Criteria in discordance.

Therefore, $C_2 (C, D) = C_3 (C, D) = 1$

But, because C_1 is neither in concordance nor in discordance,

$$C_1(C, D) = \frac{p_j - (g_j(b) - g_j(a))}{p_j - q_j} = \frac{1 - (1)}{1 - 0} = 0$$

Notice that because C_1 was scored through a qualitative scale then p_1, q_1 and $g_j (b) - g_j(a)$ are assessed in "number of categories".

And because $K = 5 + 3 + 2 = 10$ the value of $C(C, D)$ will be:

$$C(C, D) = \frac{5 \times C_1(C, D) + 3 \times C_2(C, D) + 2 \times C_3(C, D)}{10} = \frac{0 + 3 + 2}{10} = 0.5$$

For $C(D, C)$, it is already known that C_1 and C_3 are both in concordance, and C_2 is in discordance. Therefore, $C_1 (D, C) = C_3 (D, C) = 1$ and $C_2 (D, C) = 0$. Which leads to:

$$C(D, C) = \frac{0 + 3 + 2}{10} = 0.7$$

Let's evaluate another pair of Outranking Relations: $C(A, B)$ and $C(B,A)$. It was shown, in the previous section, that AQ_1B, BQ_2A and BQ_3A. Therefore, for ASB: $C_1 (A, B) = 1$;

$$C_2(A, B) = \frac{2 - (-2 - (-4))}{2 - 1} = \frac{2 - 2}{1} = 0;$$

$$C_3(A, B) = \frac{12 - (84.6 - 75.4)}{12 - 6} = \frac{12 - 9.2}{6} = 0.47$$

And $\quad C(A, B) = \dfrac{5 \times 1 + 3 \times 0 + 2 \times 0.47}{10} = \dfrac{5.93}{10} = 0.593$

For BSA: $\quad C_1(B, A) = \dfrac{1 - (1)}{1 - 0} = 0;$

$$C_2(B, A) = C_3(B, A) = 1$$

And $\quad C(A, B) = \dfrac{5 \times 0 + 3 \times 1 + 2 \times 1}{10} = \dfrac{5}{10} = 0.5$

Although it is not necessary, and it is not usually implemented as an ELECTRE III's step, the authors believe that it is useful to present the individual contributions of each Criterion to Concordance level in a matrix. These matrices are presented in Table 4.10.

Table 4.10: Contributions of each Criterion for $C(a, b)$

$C_1(a, b)$	A	B	C	D
A	1	1	1	1
B	0	1	1	1
C	0	0	1	0
D	0	1	1	1

C_2(a, b)	A	B	C	D
A	1	0	1	1
B	1	1	1	1
C	1	1	1	1
D	0	0	0	1

C_3(a, b)	A	B	C	D
A	1	0.47	0	0
B	1	1	0.15	1
C	1	1	1	1
D	1	1	1	1

With all these contributions calculated it is easy to build the Concordance Matrix:

$$C(a, b) = \frac{5 \times C_1(a, b) + 3 \times C_2(a, b) + 2 \times C_3(a, b)}{10}$$

Table 4.11: Concordance Matrix

C(a, b)	A	B	C	D
A	1	0.593	0.8	0.8
B	0.5	1	0.83	1
C	0.5	0.5	1	0.5
D	0.2	0.7	0.7	1

Summarizing, the Concordance level $C(a,b)$ represent the fraction, relatively to the sum of Importance Coefficients, of the contributions to the concordance from those criteria that are not in discordance with aSb. But, as was mentioned before, the Concordance Matrix only takes in consideration the Principium of concordance...

4.3.4   Discordance Matrices

If the Concordance level $C(a, b)$ assesses the strength of aSb then the Discordance will assess the strength of the Criteria opposing the Outranking Relation. ELECTRE III introduces the concept of Veto Threshold for Criterion j, or v_j. These values may trigger the total rejection of aSb, if the Discordance of, at least, one Criterion is too strong. Obviously, for Criteria with a non-decreasing scoring function, $v_j > p_j > q_j$.

The Discordance level for Criterion C_j, D_j (a, b), is defined by a function similar to the Concordance level except that this function relates $g_j(b) - g_j(a)$ with both v_j and p_j.

$$D_j(a, b) = \begin{cases} 0 & \text{if } g_j(b) - g_j(a) \leq p_j \\ 1 & \text{if } g_j(b) - g_j(a) \leq p_j \\ \dfrac{v_j - (g_j(b) - g_j(a))}{v_j - p_j} & \text{otherwise} \end{cases}$$

Notice that $D_j(a, b)$, the opposition of Criterion C_j to the Outranking Relation aSb, is:

- 0, if and only if C_j is not in discordance with the relation, i.e., if aP_jb, or aQ_jb, or aI_jb (and bI_ja) or bQ_ja;
- > 0, if and only if C_j is in discordance with the relation, i.e., if bP_ja:
  - If the discordance is "too strong", meaning $g_j(b) - g_j(a) > v_j$, then the opposition is total: $D_j(a,b) = 1$;
  - If C_j is in discordance but not too strongly, i.e., $p_j \leq g_j(b) - g_j(a) < v_j$ then $D_j(a,b) \in [0,1]$.

Figure 4.6 presents the graphic of $D_j(a, b)$. Notice that if $p_j \leq g_j(b) - g_j(a) < v_j$ then $D_j(a, b)$ will be a linear interpolation between the points $(p_j, 0)$ and $(v_j, 1)$.

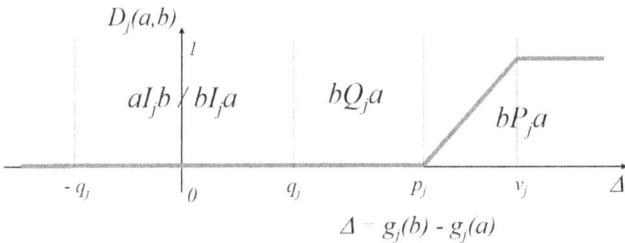

Figure 4.6: Graphic Representation of $D_j(a, b)$

Example

Let's introduce the Veto Thresholds to the Example.

Table 4.12: Example with Veto Threshold

		C_1 Interview	C_2 Demerits in Practical Test	C_3 Theoretical Test
		5	3	2
	v_j	3 categories	3 demerits	18%
	p_j	1 category	2 demerits	12%
	q_j	0 categories	1 demerits	6%
Candidates	A	Very Good	-4	75.4%
	B	Good	-2	84.6%
	C	Regular	-3	95.7%
	D	Good	-6	90.2%

For C_3, previously it was detected that: BQ_3A; CP_3 A; DP_3A; CQ_3B; DI_3B (and BI_3D); CI_3D (and DI_3C).

Therefore, $D_3(a, b) = 0$ for all pairs of Alternatives except for $D_3(A, C)$ and $D_3(A, D)$, because CP_3A and DP_3A are the only Strong Preference relations.

$$g_3(C) - g_3(A) = 95.7 - 75.4 = 20.3 \quad \text{and} \quad 20.3 > v_3 = 18 \Rightarrow D_3(A, C) = 1$$
$$g_3(D) - g_3(A) = 90.2 - 75.4 = 14.8 \quad \text{and} \quad 12 = p_3 < 14.8 \leq 18 = v_3 \Rightarrow$$

$$\Rightarrow D_3(A, D) = \frac{18 - 14.8}{18 - 12} = \frac{3.2}{6} = 0.53$$

So, the Discordance Matrix for C_3 is shown in Table 4.13.

Table 4.13: Discordance Matrix for C_3

D_3 (a, b)	A	B	C	D
A	0	0	1	0.53
B	0	0	0	0
C	0	0	0	0
D	0	0	0	0

The Discordance Matrices for C_1 and C_2 are presented in Table 4.14.

Table 4.14: Discordance Matrices for C_1 and C_2

D_1 (a, b)	A	B	C	D
A	0	0	0	0
B	0	0	0	0
C	0.5	0	0	0
D	0	0	0	0

D_2 (a, b)	A	B	C	D
A	0	0	0	0
B	0	0	0	0
C	0	0	0	0
D	0	1	0	0

Currently, for each Outranking Relation aSb there are assessments that, on one hand, measure the relation strength, and on the other hand assess each Criterion opposition to the relation. The question now is, how to aggregate all information into a single final value? The answer lies in the Credibility Degree.

4.3.5   Credibility Matrix

The final measure that assesses an Outranking Relation, aSb, is the Credibility Degree, represented is by $S(a, b)$ and is defined as:

$$S(a, b) = \begin{cases} C(a, b) & \text{if } Dj(a, b) \le C(a, b) \forall j \\ C(a, b) \times \prod\limits_{j \in J(a, b)} \dfrac{1 - D_j(a, b)}{1 - C(a, b)} & \text{otherwise} \end{cases}$$

where $J(a, b)$ is the set of all Criteria for which $D_j(a, b) > C(a, b)$.

$S(a, b)$ assesses the Credibility of the Outranking Relation aSb. If the Concordance is higher than the opposition then it is not necessary to rectify it and the Credibility will be equal to $C(a, b)$. Otherwise, the process will reduce the concordance value introducing the opposition's effect. Notice that if there is at least one Criterion C_k with $D_k(a, b) = 1$ the Credibility of aSb will be reduced to zero even if $C(a, b)$ is close to 1.

In the Example, because there are only four $D_j(a, b) > 0$ ($D_1 (C, A)$, $D_2 (D, B)$, $D_3(A, C)$ and $D_3(A, D)$) the Credibility Matrix will be equal to the Concordance Matrix except for (C, A), (D, B), (A, C) and (A, D). For these four pairs, the Concordance may need to be rectified.

For (C, A), $C(C, A) = 0.5$ and $D_1(C, A) = 0.5$. The values are equal therefore a rectification is unnecessary. But, let's see what happens if the adjustment is applied anyway:

$$S(C, A) = 0.5 \times \frac{1 - 0.5}{1 - 0.5} = 0.5 \times \frac{0.5}{0.5} = 0.5$$

But imagine that $D_1 (C, A) = 0.55$. In that case, the rectification would have been necessary:

$$S(C, A) = 0.5 \times \frac{1 - 0.55}{1 - 0.5} = 0.5 \times \frac{0.45}{0.5} = 0.45$$

and the opposition to CSA would have reduced $C(C, A)$, although just a little because the discordance is only slightly higher than the concordance.

For (D, B), $C(D, B) = 0.7$ and $D_2(D, B) = 1$. Therefore $S(D, B) = 0$. Although $C(D, B)$ is considerable high, due to the absolute veto, the credibility of DSB is zero.

For (A, C), $C(A, C) = 0.7$ and $D_3(A, C) = 1 \Rightarrow S(A, C) = 0$

Finally, for (A, D), $C(A, D) = 0.8$ and $D_3 (A, D) = 0.53$. The discordance is not enough to affect the concordance and, therefore, $S(A, D) = 0.8$

The Example Credibility Matrix is presented in Table 4.15. The concordance values rectified by the discordance are shown written in bold.

Table 4.15: Example's Credibility Matrix

S(a, b)	A	B	C	D
A	1	0.593	0	0.8
B	0.5	1	0.83	1
C	0.5	0.5	1	0.5
D	0.2	0	0.7	1

The Credibility Matrix will be used to rank the Alternatives.

4.3.6   Alternatives Ranking

As was mentioned in the previous section, with the Credibility Matrix it will be possible to rank the Alternatives. There are several methods to extract a ranking from a Credibility Matrix. One warning, though: different methods do not necessarily lead to the same ranking!

This section will present one of the most simple and commonly used methods for extracting a ranking from the Credibility Matrix. The method begins by creating two partial rankings, the Ascendant Distillation and the Descendent Distillation. The final ranking will be the intersection of both partial rankings.

The first step is to set a Minimal Value for Credibility (MVC). This value should be close to 1. But if it is too low then the final ranking it may end up with will be with several Alternatives being incomparable. If it is set too low then it may lead to a higher number of indifferences and ties.

The MVC will be used to build the matrix T. This is a binary matrix that is obtained from the Credibility Matrix by transforming all Credibility values greater or equal to the MVC into 1:

$$T(a, b) = \begin{cases} 1 & \text{if } S(a, b) \geq MVC \\ 0 & \text{otherwise} \end{cases}$$

In the example, setting MVC = 0.7.

Table 4.16: Example's Matrix T, with MVC = 0.7

T(a, b)	A	B	C	D
A	1	0	0	1
B	0	1	1	1
C	0	0	1	0
D	0	0	1	1

The T Matrix will be used to calculate the Qualification of an Alternative, $Q(a)$.

$$Q(a) = \sum_j T(a, j) - \sum_i T(a, i)$$

Or, in other words, the Qualification of an Alternative is the difference between the number of Alternatives that might be outranked by Alternative a and the number of Alternatives that might be outranking a. Alternatives with higher Qualification will appear in the top ranks, and vice-versa.

In the example:

- $Q(A) = 2 - 1 = 1$
- $Q(B) = 3 - 1 = 2$
- $Q(C) = 1 - 3 = -2$
- $Q(D) = 2 - 3 = -1$

Descendent Distillation

The first partial ranking, the Descendent Distillation, will build, iteratively, a partition of the Alternatives' set, beginning from the Alternative with the highest Qualification. The Alternatives in the first sub-set will be in the 1st rank. Then, these Alternatives are removed from matrix T and the Qualifications are recalculated. Now, the Alternatives with the highest Qualification will go to the 2nd rank, and so on until all Alternatives are ranked. When there is a tie in any of the iterations, the method tries to break it by recalculating the Qualification from the sub-matrix of T composed solely by the tied Alternatives.

Let's see the Example's Distillation: The Alternative with the highest Qualification is B. Therefore $X_1 = \{B\}$, and B will occupy the 1st rank. Removing B from the matrix T and recalculating the Qualifications:

$T(a, b)$	A	C	D
A	1	0	1
C	0	1	0
D	0	1	1

- $Q(A) = 2 - 1 = 1$
- $Q(C) = 1 - 2 = -1$
- $Q(D) = 2 - 2 = 0$

Now, the highest Qualification belongs to A, therefore $X_2 = \{A\}$, and this Alternative will go to the 2nd rank.

$T(a, b)$	C	D
C	1	0
D	1	1

Finally,

- $Q(C) = 1 - 2 = -1$
- $Q(D) = 2 - 1 = 1$

That means that $X_3 = \{D\}$ and $X_4 = \{C\}$.
The partition is: $X = X_1 \cup X_2 \cup X_3 \cup X_4 = \{B\} \cup \{A\} \cup \{D\} \cup \{C\}$.
And the Partial Ranking is:

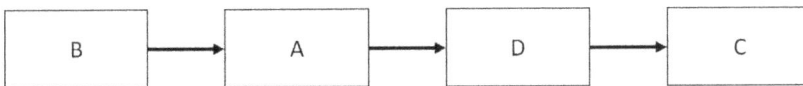

Figure 4.7: Example's Descendent Distillation

This ranking can be expressed as a set of Outranking Relations: *DescD* = $\{BSA, ASD, DSC\}$.

But sometimes, the set can include all relations generated by transitivity. That might be helpful, in the end, for the extraction of the final ranking. In this case the set would be: {**BSA, ASD, DSC**, BSD, BSC, ASC}. The bolded relations correspond to the direct ones, and the non-bolded relations to the ones obtained by transitivity.

Ascendant Distillation

This partial ranking's building is similar to the previously presented but the method begins in the bottom rank and, in each iteration, chooses the Alternative with the lowest Qualification. The ranking is built from the bottom to the top.

In the example, the Alternative with the lowest Qualification is C, therefore, C will be in the bottom rank and the first sub-set is $X_n = \{C\}$. Obviously, $n \le 4$, but due to potential ties only at the end will be possible to know how many sub-sets were necessary to build the partition.

Removing C from the matrix T.

$T(a, b)$	A	B	D
A	1	0	1
B	0	1	1
D	0	0	1

- $Q(A) = 2 - 1 = 1$
- $Q(B) = 2 - 1 = 1$
- $Q(D) = 1 - 3 = -2$

$X_{(n-1)} = \{D\}$ and D will be in the second from the bottom rank.

$T(a, b)$	A	B
A	1	0
B	0	1

- $Q(A) = 0$
- $Q(B) = 0$

The process ends up in a tie and, at this moment, it is impossible to break the tie because these are the last Alternatives in the process. Therefore the partition will be: $X = X_1 \cup X_2 \cup X_3 = \{A, B\} \cup \{D\} \cup \{C\}$.

There are, actually, two different interpretations for the ties. The first and simplest is just to assign the tied Alternatives to the same rank, therefore the partial ranking coming out from the Ascendant Distillation is:

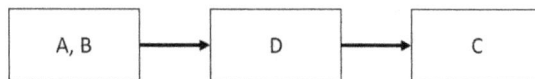

Figure 4.8: Example's Ascendant Distillation

Before presenting the other interpretation for ties, let's calculate the final ranking.

$DescD = \{BSA, ASD, DSC\}$.
$AscD = \{ASB, BSA, ASD, BSD, DSC\}$.
$DescD \cap AscD = \{BSA, ASD, DSC\}$ and the final ranking will be:

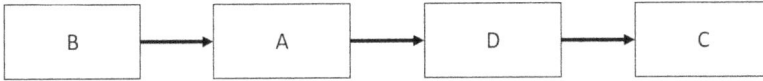

Figure 4.9: Final ranking

Let's go back to the Ascendant Distillation and see how the other interpretations for ties work.

When, in any iteration, two Alternatives end up tied two different situations may arise:

$T(a, b)$	x	y
x	1	1
y	1	1

or

$T(a, b)$	x	y
x	1	0
y	0	1

In both situations the tie is obvious: $Q(x) = Q(y) = 0$

However, in the first, both Alternatives seem to be outranking each other or: xSy and ySx. That may mean that the alternatives are indifferent and they could be placed in the same rank, corresponding to the tie illustrated in Fig. 4.10.

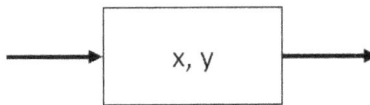

Figure 4.10: Tie with Indifference

In the second situation the alternatives do not seem to be outranking each other, or: $\sim xSy$ and $\sim ySx$ and that corresponds to an incomparabilities. That can be represented as a tie in the same rank but the differences between the Alternatives prevent them from being comparable, and the DM would probably have a hard time deciding between them. This situation is illustrated in Fig. 4.11.

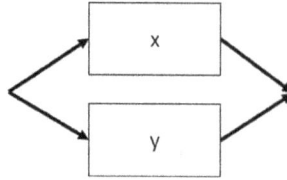

Figure 4.11: Tie with Incomparabilities

This interpretation adds an additional layer of complexity but it can make the final ranking more realistic. However, with problems with a high number of Alternatives, this might make the extraction of the final ranking too complicated.
In the example, the Ascendant Distillation would be:

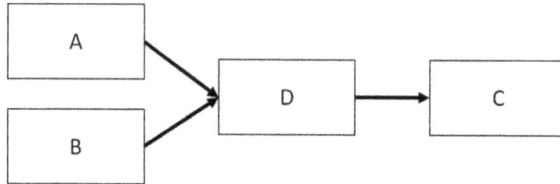

Figure 4.12: Example's Ascendant Distillation with Incomparabilities

$AscD = \{ASD, BSD, DSC\}$.
And recalling the Descendent Distillation: $DescD = \{BSA, ASD, DSC\}$.
Therefore:
$DescD \cap AscD = \{ASD, DSC\}$ which leads to the final ranking:

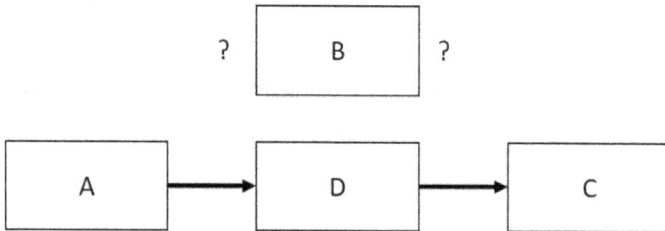

Figure 4.13: Example's Final Ranking with Incomparabilities

Using the direct Outranking Relations was not enough to rank Alternative *B* therefore it will be necessary to check the relations that can be obtained by transitivity.
$DescD = \{BSA, ASD, DSC, BSD, BSC, ASC\}$.
$AscD = \{ASD, BSD, DSC, ASC, BSC\}$.
$DescD \cap AscD = \{ASD, DSC, BSD, BSC, ASC\}$ which corresponds to:

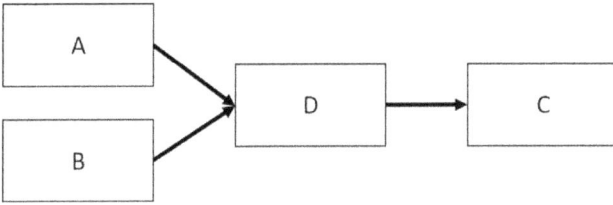

Figure 4.14: Example's Final Ranking with Incomparabilities

Notice that these rankings are very sensitive to the values chosen to be the MVC. Sometimes, small changes in this value may also lead to very different rankings.

For instance, in the example a MCV = 0.80 will lead to the following partial rankings:

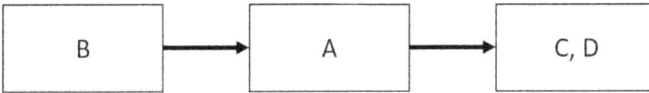

Figure 4.15: Example 2's Descendent Distillation

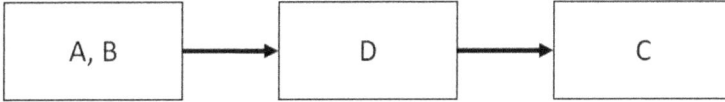

Figure 4.16: Example 2's Ascendant Distillation

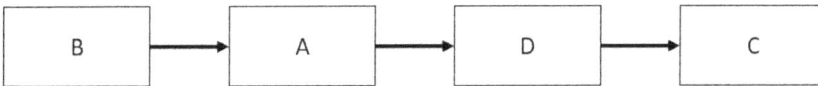

Figure 4.17: Example 2's Final Ranking

But with a MCV = 0.81:

Figure 4.18: Example 2's Descendant Distillation

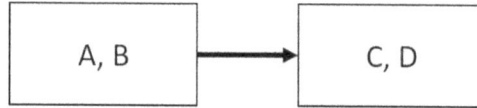

Figure 4.19: Example 2's Ascendant Distillation

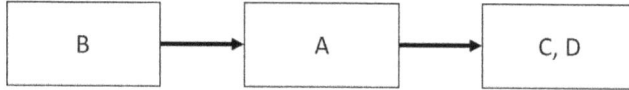

Figure 4.20: Example 2's Final Ranking

4.4 Final Remarks

ELECTRE methods incorporates the DM's preferences through the Indifference, Preference and Veto Thresholds and it can generate incredibly realistic rankings with indifference and incomparability situations. Rescaling and Value Functions are unnecessary because the method focuses on the Alternatives' score differences instead on the actual scores. And because the weights do not represent trade-off rates there is no need for a careful weight assignment which is also not fundamental. Nevertheless, as was shown, the final rankings can be complicated to extract from the Credibility Matrix and are very sensitive to small changes in the parameters. Adding to these issues is the fact that a proper sensitivity analysis is not a trivial task and, therefore, assessing the final rankings' robustness is a complicated matter.

Nevertheless, it is undeniable that the ELECTRE family of MCD methods are a very interesting non-compensatory approach to MCD problems and a perfectly viable alternative.

4.5   Proposed Exercises

1. Auntie Leonarda needs to hire a private jet plane for crossing the continent and go to an important event, tomorrow. She checked a site on-line and pre-selected the four models that pleased her most. Table 4.17 presents all the information Auntie Leonarda considers relevant for her final decision.

Table 4.17: Information Gathered by Auntie Leonarda

Car Model	Cost (in monetary units)	Trip duration (in hours)	Comfort
	C_1	C_2	C_3
Amethyst (A)	3000	6	Very Good
Beryl (B)	1500	6 h 30 min	Average
Crystal (C)	1000	10	Good
Diamond (D)	5000	4 h 15 min	Excellent

The Attribute "Comfort" was assessed by previous customers using the qualitative scale: Terrible; Very Bad; Bad; Average; Good, Very Good, Excellent.

Aiming to solve this problem using ELECTRE III, you interviewed Auntie Leonarda and settled the thresholds presented in Table 4.18.

Table 4.18: Thresholds

Threshold	Cost (in monetary units)	Trip duration (in hours)	Comfort
	C_1	C_2	C_3
Indifference	1,100	1 h	0 categories
Preference	2,200	2 h	1 category
Veto	3,300	4 h	3 categories

(a) Relatively to Criterion "Cost", indicate the existent Indifference, Weak Preference and Strong Preference Relations for each pair of Alternatives.
(b) Indicate which Criteria are in concordance and in discordance with the Relations *DSC* and *CSD*.
(c) It is a well-known fact that Auntie Leonarda considers "Comfort" the most important Criterion, followed by "Trip Duration" and that "Cost" is much less important thant the other two Criteria.
   (i) Build the Concordance Matrix $C(a, b)$.
   (ii) Build the Discordance Matrices $D_j(a, b)$.
   (iii) Obtain the Credibility Matrix $S(a, b)$.
(d) Obtain the Final ranking from the Matrix obtained in (c - iii) with a MCV = 0.75.

2. The Government of Lusoland is considering the viability of building a third bridge across river Tajus in Lisbonia metropolitan area. There are four possible places where the bridge can be built. The four alternatives were evaluated according to three points of view: Cost (in millions of Eurons); Environmental Impact (in a scale of 0 - null to 20 - catastrophic); and Road Traffic relief (in a qualitative scale "Null, Very Weak, Weak, Average, High, Very High, Extreme").

The information concerning the evaluations was collected and is presented in Table 4.19.

Table 4.19: Information Gathered Concerning Four Possible Crossing Locations

Possible crossing	Cost	Environmental Impact	Traffic Relief
Arreiro – Lisbonia	300	9	Average
Brandão – Lisbonia	150	6	Average
Ceixal – Lisbonia	100	5	Weak
Drafaria – Lisbonia	500	18	Extreme

With the objective of solving this problem using ELECTRE III, the Ministers settled the thresholds presented in Table 4.20.

Table 4.20: Thresholds

Threshold	Cost	Environmental Impact	Traffic Relief
Indifference	100	5	0 categories
Preference	250	8	1 category
Veto	450	11	3 categories

(a) Indicate the existent Indifference, Weak Preference and Strong Preference Relations for each pair of Alternatives for each Criteria
(b) Assuming that the Criterion Cost is more important than the other two, which are similarly important, calculate:
  (i) The Concordance Matrix
  (ii) The Discordance Matrices
  (iii) the Credibility Matrix
(c) From the Credibility Matrix extract a ranking using:
  (i) MCV = 0.75
  (ii) MCV = 0.85

3. Auntie Leonarda wants to throw a tremendous party to celebrate the beginning of the Social Season. Zumélia, her niece and assistant, suggested some possible themes for the party: the Aqua Party (a traditional celebration by the artificial lake in Auntie's mansion grounds, lots of fountains and waterfalls); the Black Party (everyone dressed in elegant black clothes, dark decoration and black food and beverages); the China Party (Far East themed decoration, Oriental inspired music and clothes, Chinese food); the Divine Party (guests must dress like Greek or Egyptian Mythology characters, ethereal and grandiose decoration).

"To help you out I organized all the information in a table (Table 4.21). It considers my assessment of the four alternatives from three criteria's perspective: party cost (in monetary units), impact in the jet set society (the number of celebrities who probably will attend the party), and organization difficulty (from "Very Easy" to "Very Complex")." – mentioned Zumélia.

Table 4.21: Information Organized by Zumélia

Party	Cost	Jet Set Impact	Difficulty
Aqua	2,000	60	Complex
Black	1,200	35	Very Easy
Chinese	500	40	Easy
Divine	3,000	120	Average

Zumélia, talking with Auntie Leonarda, was also able to set the Thresholds presented in Table 4.22.

Table 4.22: Thresholds

Threshold	Cost	Jet Set Impact	Difficulty
Indifference	750	25	0 categories
Preference	1,200	50	1 category
Veto	2,000	75	2 categories

(a) Indicate the existent Indifference, Weak Preference and Strong Preference Relations for each pair of Alternatives for each Criteria
(b) Assuming that the Criterion Cost is more important that the other two, which are similarly important, calculate:
  (i) The Concordance Matrix
  (ii) The Discordance Matrices
  (iii) The Credibility Matrix
(c) From the Credibility Matrix extract a ranking using:
  (i) MCV = 0.7
  (ii) MCV = 0.8

4. Evarista, the Witch has a situation. She has to run away from her coven's town because the other witches discovered she is doing good deeds for the town folks. When she applied for the place she had no idea it was an Evil Coven! And now the other crones are out there to get her. Fortunately, she has been doing this witch's life for decades and she is prepared for these kinds of circumstances. She has several caches in different towns so she can start again somewhere else. The problem is…escaping is risky and she does not know what town she prefers to escape to. After she gathered her thoughts she settled the following 6 towns that were assessed according to Cache saved, probability of getting caught and town's pleasantness (Table 4.23).

Table 4.23: Information Collected by Evarista Concerning Possible Escaping Towns

Town	Cache (in Gold pieces)	Probability of getting caught	Town's Pleasantness
Altamira	700	0.05	High
Bayona	1 500	0.25	Low
Cantlecaster	1 300	0.25	Null
Doh	240	0.00	Average
Eboricum	1 600	0.55	High
Ferim	500	0.10	High

Evarista does not want to take too high a risks. She will not try to escape to a town with a probability of being caught greater than 30%. The pleasantness Attribute was measured in the qualitative scale (Null, Very Low, Low, Average, High, Very High).

(a) Propose a set Importance Coefficients for the three Criteria and adequate Thresholds for this problem.

(b) Using the values proposed in a) built the Concordance Matrix and the Discordance Matrices
(c) Obtain the Credibility Matrix.
(d) Extract the Final Ranking using different MCV. Compare the obtained rankings and comment on them.

References

Rogers, M., and M. Bruen. (1998). Choosing realistic values of indifference, preference and veto thresholds for use with environmental criteria within ELECTRE. *European Journal of Operational Research*, 107(3), 542–551. https://doi.org/10.1016/S0377-2217(97)00175-6

Roy, B. (1968). Classement et choix en présence de points de vue multiples. RAIRO - *Operations Research - Recherche Opérationnelle*, 2(V1), 57–75.

Roy, B. (1978). ELECTRE III: Un algorithme de classements fondé sur une représentation floue des préférences en présence de critères multiples.

Analytic Hierarchy Process

5.1 Introduction

The Analytic Hierarchy Process (AHP) is a method proposed by Thomas L. Saaty during the late 1970's (T. L. Saaty 1977) and (T. Saaty 1980). The Method approaches every MCD problem by establishing different hierarchical levels and analyzing each level separately. Afterwards all levels are integrated in a single priority vector that can be used to rank the evaluated alternatives.

Figure 5.1 illustrates a typical AHP hierarchy.

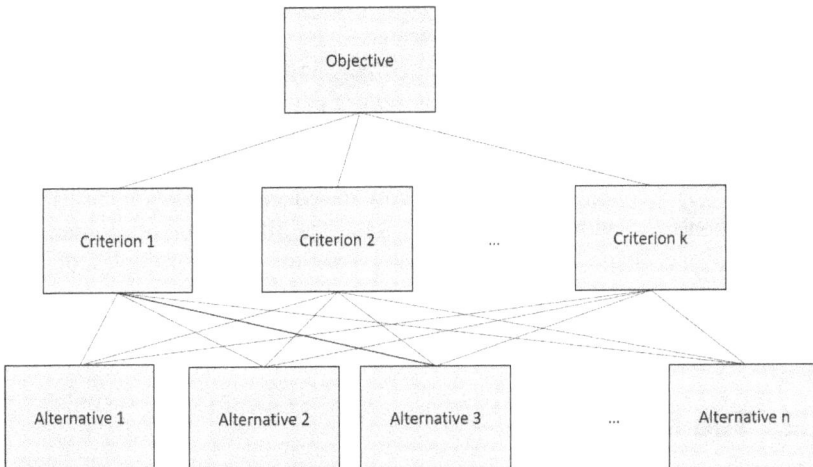

Figure 5.1: The Three Hierarchy Levels AHP Model

The method considers the Alternatives to be in the hierarchy's bottom level and the main goal of the problem's, which usually is the choice of the most adequate alternative according to the DM's points or view, to be the top level. The simplest model assumes the hierarchy has only one middle level that includes the previously established criteria. It is possible to consider more complex models that consider additional hierarchic levels, by using sub-criteria or including more than one DM.

An AHP hierarchy can be seen as a graph where all relevant entities: alternatives; criteria; and the problem's main goal, define the graph's nodes. The graph's edges represent the hierarchic relations between objects from two contiguous levels. Therefore, there are no edges either between objects at the same level nor between an object on the top and an object at the bottom level. For each node/object the AHP method will calculate a priority or weight. Priorities are values between 0 and 1 and represent the relative importance of each object when compared to the rest of the objects at the same level, according to one of the objects from the hierarchy's above level.

In each hierarchic level AHP evaluates its elements by pairwise comparisons, considering individually the perspective of each element from the above level. For instance, on the bottom level, each pair of alternatives will be submitted to several binary comparisons, one for each criterion. In the middle level, for each pair of criteria there will be a binary comparison considering the top level, which is the problem's goal. The Method then uses these pairwise comparisons to calculate the priority of a node and to establish rankings of all elements from that specific level, according to the elements of the level.

These pairwise comparisons are the kernel of AHP. They use a numerical but qualitative scale that is very intuitive and easy to use by the DM. Therefore, the method allows a relatively easy way of comparing, evaluating and ranking objects that could be considered hard to compare, or even incomparable, in compensatory and non-compensatory approaches presented in the previous Chapters.

This chapter is organized in three parts. The first one addresses the AHP method. Section 5.2.1 introduces the concepts of pairwise comparisons, Judgment Matrix and Priority Vector. The definition and importance of Consistency is then explored (section 5.2.2). The methodology is then extended to the Alternatives (section 5.2.3). Section 5.2.4 describes how to reduce a matrix's inconsistency. Then, section 5.2.5 addresses how to integrate the information of all hierarchic levels into one single vector. And finally (section 5.2.6) examples of more complex hierarchies are presented. The method's criticisms and drawbacks are explored in section 5.3. And the chapter ends with a set of exercises addressing the topics covered (section 5.4).

5.2   The Method

As mentioned before the method's core lies in the pairwise comparisons made at each hierarchy's level. These comparisons, although potentially numerous, are very easily understood by the DM contrary to some of the evaluations implemented throughout the methods presented in the previous chapters. Additionally the same procedure is used for every level of the AHP hierarchy, independent of the problem's complexity and corresponding hierarchy. In the authors' opinion these two factors are probably the main reasons behind AHP's success when applied to real life situations.

5.2.1 Pairwise Comparisons, Judgement Matrix and Priority Vector

Let's consider the middle hierarchic level for the simplest model presented above. The DM will be asked to evaluate each pair of criteria considering the qualitative scale presented below. As it can be seen below, the scale uses the odd integers from 1 to 9. The utility of the even values will be discussed later in one of the following sections.

In a problem with k criteria this analysis will correspond to $C_2^k = \dfrac{k!}{2(k-2)!}$

comparisons. These comparisons will then be collected in a matrix that is denominated the Judgment Matrix for the criteria. This matrix will be later used to calculate the Priority Vector with the criteria's weights.

The following will consider the example presented in the Appendix, with the three Criteria denominated C_1, C_2 and C_3 for simplicity.

AHP method uses the following numeric qualitative scale for the aforementioned binary comparisons:

Table 5.1: AHP Judgement Scale

1 –	Object a and b have the same interest / importance / agreeability
3 –	Object a is slightly more interesting / important / agreeable than b
5 –	Object a is more interesting / important / agreeable than b
7 –	Object a is much more interesting / important / agreeable than b
9 –	Object a is absolutely more interesting / important / agreeable than b

The even values will also be used in specific situations that will be addressed in the next sections.

Let's imagine the DM considers C_2 to be much more important than C_1 and also that C_2 is just slightly more important than C_3. So, the judgment matrix, for this DM is:

$$A = \begin{bmatrix} 1 & & \\ 7 & 1 & 3 \\ & & 1 \end{bmatrix}$$

The Judgement Matrix of a problem with k criteria will always be a square matrix of order k with all positions of its main diagonal equal to unity. The a_{ij} value corresponds to DM's comparison of criterion C_i and C_j, according to the AHP Scale, when DM considers C_i to be more important than (or has an equal importance as) C_j.

For the last comparison, between C_1 and C_3, the DM considers C_3 to be simply more important than C_1, meaning a **5** should be added to the matrix at position (3,1).

$$A = \begin{bmatrix} 1 & & \\ 7 & 1 & 3 \\ 5 & & 1 \end{bmatrix}$$

Now it will be necessary to fill out the rest of the matrix. So, what should go in the positions that are symmetrical to the ones already filled? The answer lies in one kind of matrices that have very interesting properties that AHP will explore later, the Positive Reciprocal Matrices.

A square matrix $A = [a_{ij}]$ of order k with positive entries is said to be a positive reciprocal if:

$$a_{ji} = \frac{1}{a_{ij}}, \forall i, j = 1, 2, ...k.$$

Obviously, $a_{ii} = 1, \forall i = 1, 2, ...k.$

Consequently, if we want A to be a positive reciprocal matrix, the missing entries should be the inverse of the corresponding symmetrical positions:

$$A = \begin{bmatrix} 1 & 1/7 & 1/5 \\ 7 & 1 & 3 \\ 5 & 1/3 & 1 \end{bmatrix}$$

From this Judgement Matrix, built focusing on the pairwise criteria subjective comparisons it should be possible to extract the weights of each criterion, or the criteria priority vector, according to the DM subjectivity. Because matrix A is a Positive Reciprocal Matrix it will admit a positive real eigenvalue (Bebiano et al. 2020) that will be its highest eigenvalue, λ_{max} and the method proposes the standardized eigenvector corresponding to λ_{max} as the priority vector for matrix A. Actually, for any given Judgement Matrix, this particular eigenvalue will have an important role that will be later explored and its corresponding standardized eigenvector will always be the matrix's priority vector.

As any person working with Matrix Algebra can attest, calculating the eigenvalues and their associated eigenvectors may be a cumbersome task. Fortunately, there are several numerical methods that allow fast and robust approximations to eigenvectors. The original AHP adopted two methods that will be explained below. Both methods calculate estimates for the priority vector of a given Judgment Matrix, meaning they both estimate the standardized eigenvector associated with the highest eigenvalue λ_{max}.

Method 1:

This method estimates the priority vector of a Judgement Matrix $A = [a_{ij}]$ with order n, by averaging, for each row, the elements of the standardized columns of matrix A:

$$[v_i] = \left[\frac{a'_{i1} + a'_{i2} + ... + a'_{in}}{n} \right]$$

where $a'_{ij} = \dfrac{a_{ij}}{a_{1j} + a_{2j} + ... + a_{nj}}$

Let's estimate the priority vector for

$$A = \begin{bmatrix} 1 & 1/7 & 1/5 \\ 7 & 1 & 3 \\ 5 & 1/3 & 1 \end{bmatrix}$$

Step 1, sum the elements of each column:

$$[S_1 \quad S_2 \quad S_3] = \left[1+7+5 \quad \frac{1}{7}+1+\frac{1}{3} \quad \frac{1}{5}+3+1 \right] = [13 \quad 1.476 \quad 4.2]$$

Step 2, for each column, divide each element by the corresponding sum, to obtain matrix A'.

$$A' = \begin{bmatrix} \dfrac{1}{S_1} & \dfrac{\frac{1}{7}}{S_2} & \dfrac{\frac{1}{5}}{S_3} \\ \dfrac{7}{S_1} & \dfrac{1}{S_2} & \dfrac{3}{S_3} \\ \dfrac{5}{S_1} & \dfrac{\frac{1}{3}}{S_2} & \dfrac{1}{S_3} \end{bmatrix} = \begin{bmatrix} \dfrac{1}{13} & \dfrac{1/7}{1.476} & \dfrac{1/5}{4.2} \\ \dfrac{7}{13} & \dfrac{1}{1.476} & \dfrac{3}{4.2} \\ \dfrac{5}{13} & \dfrac{1/3}{1.476} & \dfrac{1}{4.2} \end{bmatrix} = \begin{bmatrix} 0.077 & 0.097 & 0.048 \\ 0.538 & 0.677 & 0.714 \\ 0.385 & 0.226 & 0.238 \end{bmatrix}$$

As can be easily checked, for Matrix A' the sum of all elements of a given column is approximately 1.

Step 3, average the elements of each row. This will be v, the priority vector of Matrix A.

$$v = \begin{bmatrix} \dfrac{0.077 + 0.097 + 0.048}{3} \\ \dfrac{0.538 + 0.677 + 0.714}{3} \\ \dfrac{0.385 + 0.226 + 0.238}{3} \end{bmatrix} = \begin{bmatrix} 0.074 \\ 0.643 \\ 0.283 \end{bmatrix}$$

This vector indicates that, for the DM, criterion C_2 is the most important with a relative degree of importance of 64.3%. Followed by criterion C_3, with a degree of importance of 28.3%. Finally, the least important criterion, C_1, has just a degree of relative importance of 7.4%

Method 2:

The second method proposed as the priority vector of a Judgement Matrix $A = [a_{ij}]$ with order n, as the standardization of the vector v', obtain by the geometric mean of the elements of each row:

$$[v_i] = \left[\frac{v_i'}{v_1' + v_2' + \ldots + v_n'} \right]$$

where $v_i' = \sqrt[n]{a_{i1} \times a_{i2} \times \ldots \times a_{in}}$

Let's consider again matrix

$$A = \begin{bmatrix} 1 & 1/7 & 1/5 \\ 7 & 1 & 3 \\ 5 & 1/3 & 1 \end{bmatrix}$$

Step 1: Calculate v', the vector with the Geometric Means of the elements of each row:

$$v' = \begin{bmatrix} \sqrt[3]{1 \times 1/7 \times 1/5} \\ \sqrt[3]{7 \times 1 \times 3} \\ \sqrt[3]{5 \times 1/3 \times 1} \end{bmatrix} = \begin{bmatrix} \sqrt[3]{1/35} \\ \sqrt[3]{21} \\ \sqrt[3]{5/3} \end{bmatrix} = \begin{bmatrix} 0.306 \\ 2.759 \\ 1.186 \end{bmatrix}$$

Step 2: Standardize v' to obtain v:
The sum of the elements of v' is: $0.306 + 2.759 + 1.186 = 4.251$

$$v = \begin{bmatrix} \dfrac{0.306}{4.251} \\ \dfrac{2.759}{4.251} \\ \dfrac{1.186}{4.251} \end{bmatrix} = \begin{bmatrix} 0.072 \\ 0.649 \\ 0.279 \end{bmatrix}$$

The vectors obtained by the two methods are quite similar. But are any of them a good estimate of the intended vector? If v is a good approximation to a eigenvector associated to λ_{max} then $Av \approx \lambda_{max} v$.

Consequently, dividing each element of vector Av by the corresponding element of vector v we should have similar values for all of those quotients and they can be used to approximate the real eigenvalue λ_{max}.

Let's consider the vector v obtained from Method 1:

$$Av = \begin{bmatrix} 1 & 1/7 & 1/5 \\ 7 & 1 & 3 \\ 5 & 1/3 & 1 \end{bmatrix} \begin{bmatrix} 0.074 \\ 0.643 \\ 0.283 \end{bmatrix} = \begin{bmatrix} 0.222 \\ 2.008 \\ 0.886 \end{bmatrix}$$

Dividing each element of Av by the corresponding element of v:

$$\lambda_{max} \approx \frac{0.222}{0.074} \approx \frac{2.008}{0.643} \approx \frac{0.886}{0.283}$$

$$\frac{0.222}{0.074} = 3.016; \quad \frac{2.008}{0.643} = 3.121; \quad \frac{0.886}{0.283} = 3.062$$

The values are quite similar, and their average would be a good approximation of the real eigenvalue.

$$\lambda_{max} = \frac{3.016 + 3.121 + 3.062}{3} = 3.066$$

Note that λ_{max} is close to but greater than 3. That will be an important information later.

Let's check what happens for v obtained through Method 2.

$$Av = \begin{bmatrix} 1 & 1/7 & 1/5 \\ 7 & 1 & 3 \\ 5 & 1/3 & 1 \end{bmatrix} \begin{bmatrix} 0.072 \\ 0.649 \\ 0.279 \end{bmatrix} = \begin{bmatrix} 0.220 \\ 1.989 \\ 0.855 \end{bmatrix}$$

$$\lambda_{max} \cong \frac{0.220}{0.072} \cong \frac{1.989}{0.649} \cong \frac{0.855}{0.279}$$

$$\frac{0.220}{0.072} = 3.065; \frac{1.989}{0.649} = 3.065; \cong \frac{0.855}{0.279} = 3.065$$

The three quotients were exactly equal. Consequently, $\lambda_{max} \cong 3.065$

Both v and λ_{max} given by the two methods were similar, and in fact there are no differences between using either method.

Although the methods described above generated a priority vector from the Judgment Matrix evaluating the criteria's relative importance, they will also be used for each and every matrix built for every hierarchic level. Namely, in the bottom level, where Alternatives should be compared pairwise under each criterion's point of view. But we will come back to this later.

For now, a question arises: Judgment Matrix was built by the DM making some subjective choices. So, how reliable can that matrix be? How coherent is the DM on his/her subjectivity? The next section will address these questions, and introduce the concept of Consistency.

5.2.2 Consistency of Judgement Matrices. Consistency Index. Random Consistency Index. Consistency Ratio

As was presented in the previous section, Judgement Matrices are built from a set of pairwise comparisons between the objects being ranked. Those comparisons will be highly subjective and affected by the DM's biases. One problem arises: do those biases follow rational and logical reasoning or are simply emotional and irrational DM's responses? The relation "preference" is not usually transitive, but maybe the lack of transitivity of a given judgement matrix is acceptable. How to evaluate this?

When the data are quantitative it is easy to understand the reasoning behind the judgment processes but when the problem deals with qualitative information it is easy to wonder if DM's judgments are just too irrational and non-transitive. Additionally, if the problem has a high number of alternatives it is not that hard to imagine a situation where the DM gets bored with all comparisons being asked and starts answering randomly.

The aforementioned situations can lead to judgement matrices that may not reflect a coherent opinion and, consequently, will produce a priority vector that may not represent a correct ranking from the DM's point of view. Therefore it is very important to assess the judgment matrices' lack of transitivity.

A positive reciprocal matrix $A = [a_{ij}]$ of order k is considered to be Consistent or Transitive if:

$$a_{ij} = a_{im} \times a_{mj}, \forall i, j, m = 1, 2, \ldots k$$

Some remarks about this definition:

1. Notice that, for judgement matrices of order k, if $i = j$ then $a_{ii} = a_{im} \times a_{mi} = a_{im} \times \dfrac{1}{a_{im}} = 1$ is always true for all $m = 1, 2, \ldots, k$;

2. Additionally, when $m = j$: $a_{ij} = a_{im} \times a_{mi} = a_{ij} \times a_{jj} = a_{ij} \times 1 = a_{ij}$. Similarly for when $m = i$. Therefore, to verify if a judgement matrix is consistent it will only be necessary to check the equality for the index combinations where $m \neq j \wedge m \neq i \wedge i \neq j$;

3. Finally, it is also only necessary to verify above (or below) the main diagonal, because, give i and j, if $a_{ij} = a_{im} \times a_{mi}$ for all m, then $a_{ji} = \dfrac{1}{a_{ij}} = \dfrac{1}{a_{im} \times a_{mj}} = \dfrac{1}{a_{mj}} \times \dfrac{1}{a_{im}} = a_{jm} \times a_{mi}$, for all m.

A positive reciprocal matrix that does not verify the above condition is non-consistent or, more informally and inside AHP context, inconsistent.

Most judgment matrices are not consistent but when the DM's answers to the pairwise comparisons follow a certain coherence then the corresponding judgment matrix will not have a too high level of inconsistency. The issue is: how to evaluate this?

Let's consider the previous matrix

$$A = \begin{bmatrix} 1 & 1/7 & 1/5 \\ 7 & 1 & 3 \\ 5 & 1/3 & 1 \end{bmatrix}$$

For A to be consistent then $a_{21} = 7$, should be equal to $a_{23} \times a_{31}$. But, $a_{21} = 7$, and $a_{23} \times a_{31} = 3 \times 5 = 15$. So, Matrix A is not consistent according to the definition. But is its inconsistency acceptable?

For a given consistent positive reciprocal matrix $A = [a_{ij}]$ of order k it is know that its highest eigenvalue is $\lambda_{max} = k$. For non-consistent matrices, $\lambda_{max} > k$.

These results' demonstration can be found here (Shiraishi et al. 1998).

Higher levels of inconsistency will correspond to λ_{max} being much greater than k. Therefore the difference between λ_{max} and k can be used as a measure of a matrix lack of consistency. AHP method proposes the Consistency Index (**CI**) using that difference:

$$CI = \frac{\lambda_{max} - k}{k - 1}$$

Obviously, if a judgment matrix is consistent, $CI = 0$

Nevertheless, CI on its own is not enough to assess if the matrix is too inconsistent. To evaluate that Saaty calculated the average CI for a high number of Judgment Matrices randomly generated of order greater than 2. He called that value the Random Consistency Index (*RI*).

Table 5.2 presents the RI values for matrices of order up to 15

Table 5.2: Random Consistency Index for Matrices of Order k

Order	3	4	5	6	7	8	9	10	11	12	13	14	15
RI	0.58	0.90	1.12	1.24	1.32	1.41	1.45	1.49	1.51	1.48	1.56	1.57	1.59

So, a judgment matrix has an acceptable inconsistency if its CI is relatively low when compared to the average CI of random matrices of the same order. AHP recommends that a matrix may be considered as having an acceptable inconsistency if the ratio $RI = \frac{CI}{RI}$ is lesser than 0.1. RI is also denominated as the Consistency Ratio.

For matrix A, with λ_{max} calculated by Method 2:

$$CI = \frac{3.065 - 3}{3 - 1} = \frac{0.065}{2} = 0.033$$

Matrices of order 3 have a $RI = 0.58$

Therefore, Matrix A has a Consistency Ratio of:

$$CR = \frac{0.033}{0.58} = 0.056 < 0.1$$

Although matrix A is not really consistent, its inconsistency level is sufficiently low that its corresponding priority vector v can be considered a good indicator of the DM's preferences.

An important remark: all judgement matrices of order 2 are consistent:

$$A = \begin{bmatrix} 1 & a_{12} \\ 1/a_{12} & 1 \end{bmatrix}$$

For $k = 2$ there are only eight index combinations of $a_{ij} = a_{im} \times a_{mi}$ to be verified, and all are trivial. For example:

$$a_{11} = a_{12} \times a_{21} = a_{12} \times \frac{1}{a_{21}} = 1$$

$$a_{22} = a_{21} \times a_{12} = \frac{1}{a_{12}} \times a_{12} = 1$$

Additionally, any of the methods presented in section 5.2.1 will give the same vector v and $\lambda_{max} = 2$, as expected.

Method 1:

$$\begin{bmatrix} S_1 & S_2 \end{bmatrix}\begin{bmatrix} 1+a_{21} & 1+a_{12} \end{bmatrix} = \begin{bmatrix} 1+1/a_{12} & 1+a_{12} \end{bmatrix} = \begin{bmatrix} \dfrac{1+a_{12}}{a_{12}} & 1+a_{12} \end{bmatrix}$$

$$A' = \begin{bmatrix} \dfrac{a_{12}}{1+a_{12}} & \dfrac{a_{12}}{1+a_{12}} \\ \dfrac{a_{12}/a_{12}}{1+a_{12}} & \dfrac{1}{1+a_{12}} \end{bmatrix} = \begin{bmatrix} \dfrac{a_{12}}{1+a_{12}} & \dfrac{a_{12}}{1+a_{12}} \\ \dfrac{1}{1+a_{12}} & \dfrac{1}{1+a_{12}} \end{bmatrix}$$

$$v = \begin{bmatrix} \dfrac{2a_{12}}{2(1+a_{12})} \\ \dfrac{2}{2(1+a_{12})} \end{bmatrix} = \begin{bmatrix} \dfrac{a_{12}}{1+a_{12}} \\ \dfrac{1}{1+a_{12}} \end{bmatrix}$$

Method 2:

$$v' = \begin{bmatrix} \sqrt{1 \times a_{12}} \\ \sqrt{1/a_{12} \times 1} \end{bmatrix} = \begin{bmatrix} \sqrt{a_{12}} \\ \dfrac{\sqrt{a_{12}}}{a_{12}} \end{bmatrix}$$

$$S = \sqrt{a_{12}} + \frac{\sqrt{a_{12}}}{a_{12}} = \frac{(a_{12}+1)\sqrt{a_{12}}}{a_{12}}$$

For λ_{max}: $v = \begin{bmatrix} \dfrac{a_{12}\sqrt{a_{12}}}{(a_{12}+1)\sqrt{a_{12}}} \\ \dfrac{a_{12}\sqrt{a_{12}}}{a_{12}(a_{12}+1)\sqrt{a_{12}}} \end{bmatrix} = \begin{bmatrix} \dfrac{a_{12}}{a_{12}+1} \\ \dfrac{1}{a_{12}+1} \end{bmatrix}$

$$Av = \begin{bmatrix} 1 & a_{12} \\ \dfrac{1}{a_{12}} & 1 \end{bmatrix}\begin{bmatrix} \dfrac{a_{12}}{a_{12}+1} & \dfrac{1}{a_{12}+1} \end{bmatrix} = \begin{bmatrix} \dfrac{a_{12}}{a_{12}+1} + \dfrac{a_{12}}{a_{12}+1} & \dfrac{a_{12}}{a_{12}(a_{12}+1)} + \dfrac{1}{a_{12}+1} \end{bmatrix}$$

$$= \begin{bmatrix} \dfrac{2a_{12}}{a_{12}+1} & \dfrac{2}{a_{12}+1} \end{bmatrix}$$

Dividing each element of Av by the corresponding element of v:

$$\frac{\dfrac{2a_{12}}{a_{12}+1}}{\dfrac{a_{12}}{a_{12}+1}} = 2;$$

$$\frac{2}{\dfrac{a_{12}+1}{\dfrac{1}{a_{12}+1}}} = 2$$

Therefore, $\lambda_{max} = 2$. And, consequently, $CI = 0$

5.2.3   Concerning the Alternatives

In the previous sections the AHP method was presented using the interaction between the top and the middle hierarchic level of an AHP hierarchy as an example. But it was mentioned that the same methodology should be applied for all hierarchic levels.

How does the method work for the Bottom level, where Alternatives lie?

Middle level has three objects, Criteria C_1, C_2 and C_3. And the priority vector that defines the relative importance of the Criteria was obtained in section 5.2.1. For the Bottom level the priority vectors have to consider the interaction between Alternatives, at the bottom level, and Criteria, at the level immediately above: the middle level. Because there are three criteria, Alternatives' pairwise comparisons will generate three judgement matrices, one for each point of view.

Consider, again, the example, presented in the Appendix.

Table 5.3: Example Data

		Criteria		
		C_1	C_2	C_3
Alternatives	A	Very Good	4	75.4
	B	Good	2	84.6
	C	Regular	3	95.7
	D	Good	6	90.2

"For criterion C_1 – Interview, the DM shall compare each pair of possible "Interview Rates" showed in Table 5.3 and assign a score to each comparison, using the AHP scale presented in Table 5.1. The possible Interview rates are: "Very Good" (awarded to Candidate A), "Good" (awarded to Candidates B and D) and, finally, "Regular" (awarded to Candidate C). Therefore, the following comparisons should be assessed by the DM: "Very Good" versus "Good"; "Very Good" versus "Regular"; and "Good" *versus* "Regular".

Assume that DM considered a result of "Very Good" in the interview was more interesting than a rate of "Good" (corresponding to a score of 5 in the AHP scale) and that "Very Good" is much more interesting than a rate of "Regular" (7 in the AHP scale). Additionally, a result of "Good" was slightly more interesting than a result of "Average" (score 3).

That means that the judgment matrix concerning the Alternatives, considering criterion C_1's point of view will be:

$$A_1 = \begin{bmatrix} 1 & 5 & 7 & 5 \\ 1/5 & 1 & 3 & 1 \\ 1/7 & 1/3 & 1 & 1/3 \\ 1/5 & 1 & 3 & 1 \end{bmatrix}$$

Notice that $a_{24} = a_{42} = 1$ because both Candidates B and D had the result of "Good" in the interview.

This Matrix will correspond to:

$$v_1 = \begin{bmatrix} 0.632 \\ 0.153 \\ 0.061 \\ 0.153 \end{bmatrix}, \lambda_{max} = 4.073, \ CI = \frac{4.0733 - 4}{4 - 1} = 0.024,$$

$RI = 0.9$, (Table 4.2), $CI = \dfrac{0.024}{0.9} = 0.027$

Vector v_1 represents the relative interest DM feels about the four candidates, considering solely their evaluation during the interviews. The vector v_1 can be used because matrix A_1 has an acceptable level of inconsistency ($CI < 0.1$).

The process should be repeated for the other two criteria...

For criterion C_2 the pairwise comparisons resulted on the following judgement matrix and vector:

$$A_2 = \begin{bmatrix} 1 & 1/5 & 1/3 & 5 \\ 5 & 1 & 3 & 7 \\ 3 & 1/3 & 1 & 5 \\ 1/5 & 1/7 & 1/5 & 1 \end{bmatrix}, v_2 = \begin{bmatrix} 0.133 \\ 0.559 \\ 0.261 \\ 0.048 \end{bmatrix}$$

with the corresponding values: $\lambda_{max} = 4.240$, $CI = 0.080$, $RI = 0.9$, $CI = 0.089$.

Again, vector v_2 is reliable because matrix A_2 has $CI < 0.1$

For criterion C_3 the pairwise comparisons resulted on the following:

$$A_3 = \begin{bmatrix} 1 & 1/5 & 1/7 & 1/7 \\ 5 & 1 & 1/5 & 1/5 \\ 7 & 5 & 1 & 3 \\ 7 & 5 & 1/3 & 1 \end{bmatrix}, v_3 = \begin{bmatrix} 0.042 \\ 0.112 \\ 0.536 \\ 0.310 \end{bmatrix}$$

$\lambda_{max} = 4.364$, $CI = 0.121$, $RI = 0.9$, $CR = 0.135$.

Unfortunately, vector v_3 is not that reliable because matrix A_3 has $CR = 0.135 > 0.1$.

But, how problematic can this be? This situation will be explored in the next section. For now, admit that is not a relevant problem.

Now that the method was applied to all hierarchic levels we can review all the priority vectors:

$$v = \begin{bmatrix} 0.072 \\ 0.649 \\ 0.279 \end{bmatrix} \begin{matrix} C1 \\ C2, \\ C3 \end{matrix} \qquad v_1 = \begin{bmatrix} 0.632 \\ 0.153 \\ 0.061 \\ 0.153 \end{bmatrix}, \quad v_2 = \begin{bmatrix} 0.133 \\ 0.559 \\ 0.261 \\ 0.048 \end{bmatrix}, \quad v_3 = \begin{bmatrix} 0.042 \\ 0.112 \\ 0.536 \\ 0.310 \end{bmatrix} \begin{matrix} A \\ B \\ C \\ D \end{matrix}'$$

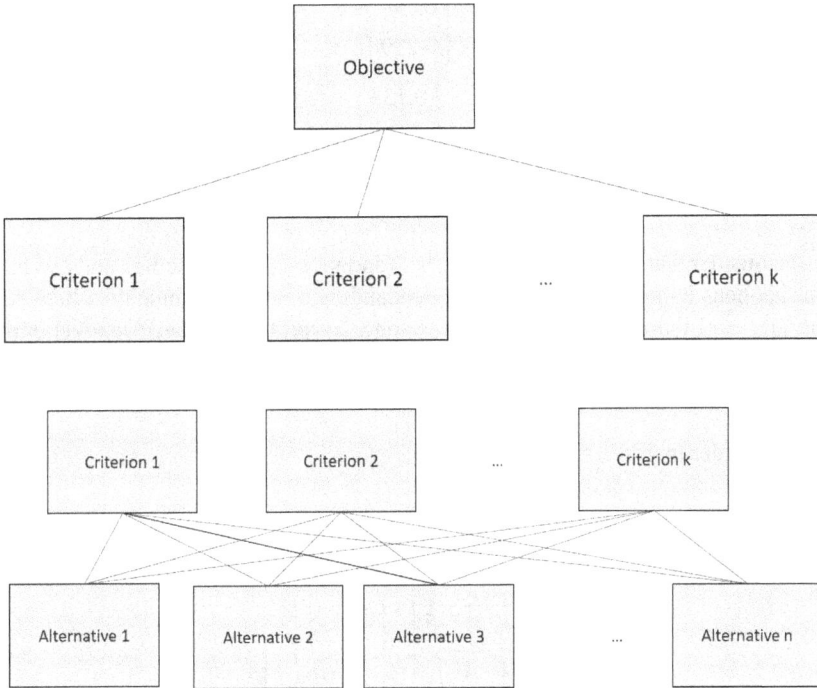

These vectors present the relative importance of each Criterion for the DM and the relative interest of each Alternative, considering each Criterion, according to the DM's opinion. But how to integrate all this information? This will be explained on 5.2.5.

But, before that there is an important issue still waiting to be addressed... What if a matrix's inconsistency is too high?

5.2.4   Non-Consistency

For reasons already mentioned, sometimes a judgment matrix has RI greater than 0.1 and when that situation arises, in most situations it is recommended that the corresponding priority vector should not be used.

But what should be done when this happens?

Firstly, locate the culprits, meaning, the a_{ij} elements that differ most from the $a_{ik} \times a_{kj}$ products. The larger the difference the more it contributes to the inconsistency. Sometimes, for matrices with lower orders, this is easily done by simple observation. For larger orders it can be complicated without systematically verifying all index combinations but that is a simple and quick computational procedure. And, sometimes, the identification of those sources of inconsistency will provide an insight on how to reduce the CR.

$$v = \begin{bmatrix} 0.072 \\ 0.649 \\ 0.279 \end{bmatrix} \begin{matrix} C1 \\ C2 \\ C3 \end{matrix}$$

$$v_1 = \begin{bmatrix} 0.632 \\ 0.153 \\ 0.061 \\ 0.153 \end{bmatrix}, \quad v_2 = \begin{bmatrix} 0.133 \\ 0.559 \\ 0.261 \\ 0.048 \end{bmatrix}, \quad v_3 = \begin{bmatrix} 0.042 \\ 0.112 \\ 0.536 \\ 0.310 \end{bmatrix} \begin{matrix} A \\ B \\ C \\ D \end{matrix}$$

Consider that $a_{ij} > 1$, and $a_{im} \times a_{mj} < 1$ (and vice-versa) for some i and j. When this happens it may point out a misunderstanding or miscommunication between the DM and the analyst. The values a_{ij} and a_{ji} might have been inadvertently swapped, causing the inconsistency.

$$\text{Consider matrix } A = \begin{bmatrix} 1 & 1/7 & 1/5 \\ 7 & 1 & 1/3 \\ 5 & 3 & 1 \end{bmatrix}.$$

Using Method 2:

$$v' = \begin{bmatrix} \sqrt[3]{1 \times 1/7 \times 1/5} \\ \sqrt[3]{7 \times 1 \times 1/3} \\ \sqrt[3]{5 \times 3 \times 1} \end{bmatrix} = \begin{bmatrix} \sqrt[3]{1/35} \\ \sqrt[3]{7/3} \\ \sqrt[3]{15} \end{bmatrix} = \begin{bmatrix} 0.306 \\ 1.326 \\ 2.466 \end{bmatrix}$$

$$S = 0.306 + 1.326 + 2.466 = 4.098$$

$$v = \begin{bmatrix} \dfrac{0.306}{4.098} \\ \dfrac{1.326}{4.098} \\ \dfrac{2.466}{4.098} \end{bmatrix} = \begin{bmatrix} 0.075 \\ 0.324 \\ 0.620 \end{bmatrix}$$

$$Av = \begin{bmatrix} 1 & 1/7 & 1/5 \\ 7 & 1 & 1/3 \\ 5 & 3 & 1 \end{bmatrix} = \begin{bmatrix} 0.075 \\ 0.324 \\ 0.620 \end{bmatrix} = \begin{bmatrix} 0.241 \\ 1.046 \\ 1.946 \end{bmatrix}$$

Dividing each element of Av by the corresponding element of v:

$$\frac{0.241}{0.075} = 3.233; \quad \frac{1.046}{0.324} = 3.233; \quad \frac{1.946}{0.620} = 3.233$$

$$\lambda_{max} = 3.233, \, CI = \frac{3.233 - 3}{3 - 1} = \frac{0.233}{2} = 0.117$$

Matrices of order 3 have a $RI = 0.58$

Therefore, matrix A has a Consistency Ratio of:

$$CR = \frac{0.117}{0.58} = 0.201 > 0.1 !!!$$

The matrix has an unacceptable inconsistency level...

Comparing a_{ij} with $a_{im} \times a_{mj}$, for the non-trivial cases and considering only the $a_{ij} > 1$:

$$a_{21} = 7 \qquad a_{23} \times a_{31} = 3 \times 5 = 15$$

$$a_{31} = 5 \qquad a_{32} \times a_{21} = 3 \times 7 = 21$$

$$a_{32} = 3 \qquad a_{31} \times a_{12} = 5 \times \frac{1}{7} = \frac{5}{7} = 0.714$$

The highest difference is in a_{31}... and that might point to the most important source of inconsistency. Although, there is one detail: $a_{32} = 3 > 1$ and $a_{31} \times a_{12} = 0.714 < 1$.

Note that:

$$a_{23} = \frac{1}{3} \simeq 0.333 < 1 \quad a_{21} \times a_{13} = 7 \times \frac{1}{5} = \frac{7}{5} = 1.4 > 1 \text{ so this issue also happens}$$

for the symmetrical case.

That may point for a swap between the values of a_{23} and a_{32}. "Reading" the pairwise comparisons: object 2 is much more important than object 1 ($a_{21} = 7$), and object 3 is more important than object 1 ($a_{31} = 7$), consequently one would expect that object 2 should be, at least, slightly more important than 3. But the information in the matrix contradicts this ($a_{23} = 1/3$)! Probably, the DM's intentions were to attribute 3 to a_{23} but some miscommunication led to the inverse being attributed.

Let's see what happen when the swap is inverted:

$$\text{Now, } A = \begin{bmatrix} 1 & 1/7 & 1/5 \\ 7 & 1 & 3 \\ 5 & 1/3 & 1 \end{bmatrix} \text{ was a previously used matrix!}$$

Its priority vector is $v = \begin{bmatrix} 0.072 \\ 0.649 \\ 0.279 \end{bmatrix}$, with a $\lambda_{max} = 3.065$, $CI = 0.033$, and $CR = 0.056 < 0.1$.

Fortunately, the situation described above does not occur that often. Most inconsistency comes not from mistakes but from the weak transitivity of the "preferable to" relation. Consider the following matrix:

$$\text{Consider matrix } A = \begin{bmatrix} 1 & 1/5 & 1/3 \\ 5 & 1 & 5 \\ 3 & 1/5 & 1 \end{bmatrix}.$$

Using Method 2:

$$v' = \begin{bmatrix} \sqrt[3]{1/15} \\ \sqrt[3]{25} \\ \sqrt[3]{3/5} \end{bmatrix} = \begin{bmatrix} 0.405 \\ 2.924 \\ 0.843 \end{bmatrix}$$

$$S = 0.405 + 2.924 + 0.843 = 4.177$$

$$v = \begin{bmatrix} 0.097 \\ 0.701 \\ 0.202 \end{bmatrix} Av = \begin{bmatrix} 0.305 \\ 2.197 \\ 0.634 \end{bmatrix}$$

Dividing each element of Av by the corresponding element of v:

$$\frac{0.305}{0.097} = 3.156; \frac{2.197}{0.701} = 3.156; \frac{0.634}{0.202} = 3.156$$

$$\lambda_{max} = 3.156, CI = \frac{3.156 - 3}{3 - 1} = 0.068$$

Matrices of order 3 have a $RI = 0.58$
Therefore, matrix A has a Consistency Ratio of:

$$CR = \frac{0.068}{0.58} = 0.117 > 0.1 !!!$$

The matrix has an unacceptable inconsistency level...
Comparing a_{ij} with $a_{im} \times a_{mj}$, for the non-trivial cases and considering only $a_{ij} > 1$:

$$a_{21} = 5 \qquad\qquad a_{23} \times a_{31} = 5 \times 3 = 15$$

$$a_{23} = 5 \qquad\qquad a_{21} \times a_{13} = 5 \times \frac{1}{3} = \frac{5}{3} = 1.667$$

$$a_{31} = 3 \qquad\qquad a_{32} \times a_{21} = 1/5 \times 5 = 1$$

The highest difference is on a_{21} (15–5 = 10). Consequently, to lower the matrix's inconsistency either the value of a_{21} should be increased or the values of a_{23} or a_{31} should be decreased. That will lower the mentioned difference and correspondingly, the inconsistency... AHP recommends the use of the even number immediately above or below the values being changed. It must be recalled that those even numbers were opportunely left off the AHP judgement scale.

Translating the values to the AHP judgement scale: object 3 is slightly more interesting than object 1 ($a_{31} = 3$). And Object 2 is more interesting than object

3 (a_{23} = 5). So, one would expect that object 2 should have a higher "degree" of interest than "more interesting" when compared to object 1... But the DM stated that object 2 is simply "more interesting" than 1, Well, probably because that higher level of interest is not enough, for the DM, to turn the relation into "much more interesting"... Thus it is natural that the intermediate even values should be used to "smooth" the judgment matrix's entries.

On this example the possibilities of correction are: $a_{21} \rightarrow 6$, $a_{23} \rightarrow 4$, and/or $a_{31} \rightarrow 2$.

Case 1: $a_{21} \rightarrow 6$

$$A = \begin{bmatrix} 1 & 1/5 & 1/3 \\ 6 & 1 & 5 \\ 3 & 1/6 & 1 \end{bmatrix}, v = \begin{bmatrix} 0.088 \\ 0.717 \\ 0.195 \end{bmatrix}, \lambda_{max} = 3.094, \; CI = 0.047, \; CR = 0.081.$$

Case 2: $a_{23} \rightarrow 4$

$$A = \begin{bmatrix} 1 & 1/5 & 1/3 \\ 6 & 1 & 5 \\ 3 & 1/6 & 1 \end{bmatrix}, v = \begin{bmatrix} 0.088 \\ 0.717 \\ 0.195 \end{bmatrix}, \lambda_{max} = 3.086, \; CI = 0.043, \; CR = 0.074.$$

Case 3: $a_{31} \rightarrow 2$

$$A = \begin{bmatrix} 1 & 1/5 & 1/2 \\ 5 & 1 & 5 \\ 2 & 1/5 & 1 \end{bmatrix}, v = \begin{bmatrix} 0.113 \\ 0.709 \\ 0.179 \end{bmatrix}, \lambda_{max} = 3.054, \; CI = 0.027, \; CR = 0.046.$$

As can be seen above, any of the proposed corrections will be enough to decrease the Consistency Ratio into an acceptable level.

Some remarks: unfortunately, it is not always that easy to reduce the matrices inconsistency. Sometimes, several corrections are in order. And in the worst cases an additional consultation with the DM is necessary with the consequent total remaking of the prevaricator judgment matrix. Also, the processes described in this section are just the simplest ways of reducing inconsistency. There are more elaborate and complex mechanisms that can be found by the reader in the literature. (Shiraishi et al. 1998). Finally, under certain circumstances, having a judgement matrix with a CR slightly over 0.1 is not that much of a problem. This will be elaborated in the next section.

5.2.5   Integrating the Problem's Hierarchic Levels

Sections 5.2.2 and 5.2.3 present the priority vectors obtained for the different hierarchical levels. The main question arises now: How to integrate all those vectors into a single synthesis evaluation?

Actually, a synthesis evaluation is really simply to obtain from the priority vectors. Using a simple weighted sum that considers the elements from the top level priority vector as the weights and the elements from the bottom level vectors as the weighted terms being summed.

Matricially that weighted sum can be represented as:

$$v_{Global} = [v_1 \; v_2 \; v_3] \times v$$

Let's remind the priority vectors previously obtained:

$$v = \begin{bmatrix} 0.072 \\ 0.649 \\ 0.279 \end{bmatrix} \begin{matrix} C_1 \\ C_2 \\ C_3 \end{matrix}, \quad v_1 \begin{bmatrix} 0.632 \\ 0.153 \\ 0.061 \\ 0.153 \end{bmatrix}, v_2 \begin{bmatrix} 0.133 \\ 0.559 \\ 0.261 \\ 0.048 \end{bmatrix}, v_3 = \begin{bmatrix} 0.042 \\ 0.112 \\ 0.536 \\ 0.310 \end{bmatrix} \begin{matrix} A \\ B \\ C \\ D \end{matrix}$$

The Global priority vector will be:

$$v_{Global} = [v_1 \times v_2 \times v_3] = \begin{bmatrix} 0.072 \\ 0.649 \\ 0.279 \end{bmatrix} = \begin{bmatrix} 0.632 & 0.133 & 0.042 \\ 0.153 & 0.559 & 0.112 \\ 0.061 & 0.261 & 0.536 \\ 0.153 & 0.048 & 0.310 \end{bmatrix} \times \begin{bmatrix} 0.072 \\ 0.649 \\ 0.279 \end{bmatrix} = \begin{bmatrix} 0.143 \\ 0.405 \\ 0.303 \\ 0.128 \end{bmatrix} \begin{matrix} A \\ B \\ C \\ D \end{matrix}$$

The Global priority vector synthetized the DM's three points of view using the top level priority vector as the weights vector. That means, considering all criteria the DM will prefer Candidate A with a relative preference of 0.143, B with 0.405, and so on. More importantly, Candidate B is the one with the highest priority consequently will be the one who will please the DM the most.

Nevertheless, recall that all vectors, except v_3, were extracted from judgement matrices with acceptable values of inconsistency, but for v_3 the corresponding matrix's CR was greater than 0.1. So this issue must also be addressed. To evaluate if matrix A_3's excess of inconsistency is high enough to make the Global priority vector unreliable, one must calculate the Global Consistency Index. For the Top level, the CI will correspond to the value for matrix A : $CI_{TOP} = 0.033$. But, on the bottom level three matrices were constructed form the pairwise comparisons:A_1, A_2 and A_3, with corresponding CI: $CI_1 = 0.024$, $CI_2 = 0.080$ and $CI_3 = 0.121$.

The Consistency Index for the Bottom Level will be the weighted sum of the CI for the three matrices built on this level, using the above level's priority vector as the weights vector, similarly to the process used above, when the three priority vectors were synthesized into the Global vector.

$$CI_{BOT} = 0.072 \times CI_1 + 0.649 \times CI_2 + 0.279 \times CI_3 = 0.087$$

Therefore, the Global Consistency Index of the total hierarchy is:

$$CI_{GLOBAL} = CI_{TOP} + CI_{BOT} = 0.033 + 0.087 = 0.120$$

Naturally, the Global Random Index will be the sum of the Random Consistency Indices of all hierarchic levels:

$$RI_{GLOBAL} = RI_{TOP} + RI_{BOT} = 0.85 + 0.9 = 1.48$$

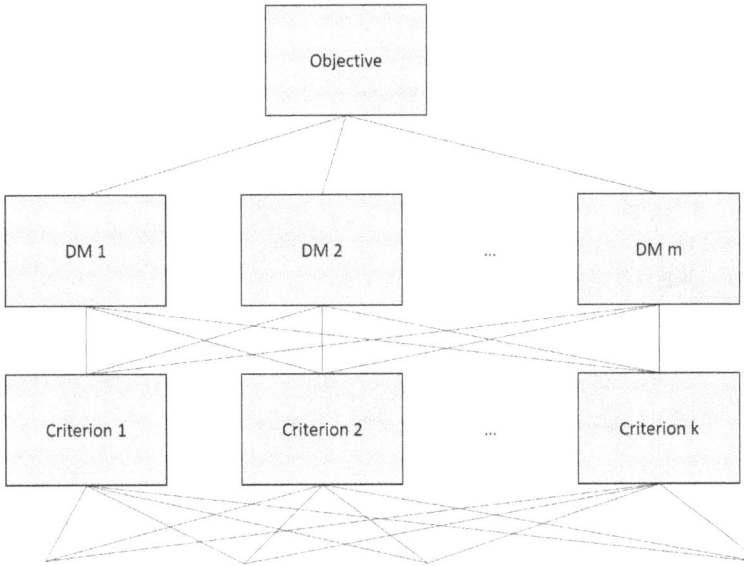

Figure 5.2: An AHP Hierarchy with Multiple Decision Makers

Remind that on the Top level the judgement matrix was of order 3 and on bottom's matrices were of order 4.

Consequently, the Global Consistency Ratio will be the quotient between *CI* and *RI*:

$$CR_{GLOBAL} = \frac{CI_{GLOBAL}}{RI_{GLOBAL}} = \frac{0.12}{0.148} = 0.08$$

In spite of the fact matrix A_3 has a *CR* greater than recommended, the overall *CR* is perfectly acceptable. That happens due to the fact that the weight of the priority vector associated with A_3 is considerably low relative to the other involved priority vectors.

5.2.6 More Complex Hierarchies

The previous sections detailed the method's procedures using the simplest of AHP hierarchies. A top level corresponding to the final goal, a middle level with the Criteria and the bottom level considering the Alternatives. Obviously some problems will need more complex hierarchies. If more than one Decision Makers are involved in the process, an additional level will be necessary, between the Top and the middle level, listing all DM. The Top Matrix will be built in the presence of all participants in the Decision Making process, and they will have to evaluate if all of them will have similar importance in the decision process or if the opinion of one of them will have more, or less weight than the others. This situation is illustrated on Fig. 5.2.

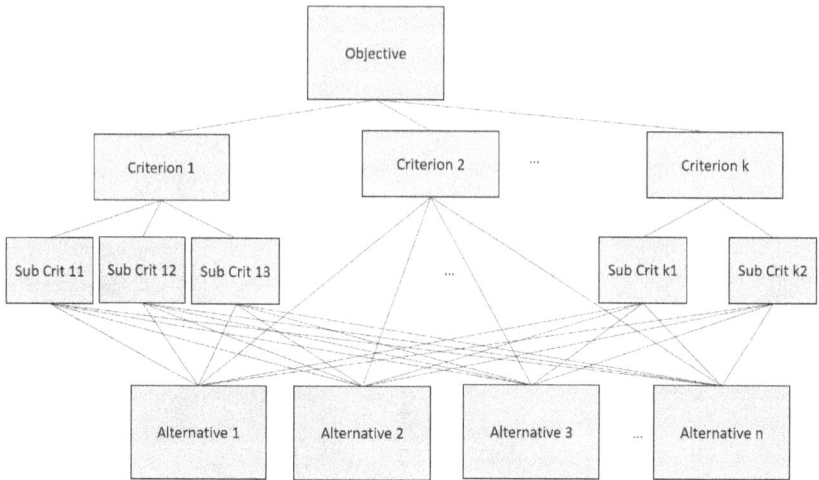

Figure 5.3: An AHP Hierarchy with Sub-criteria

In some problems the criteria might have sub-criteria making necessary the addition of another level between the bottom one and the middle level. Figure 5.3 illustrates that situation.

But independently of the hierarchical complexity, AHP procedure is always the same: a judgment matrix is built using the corresponding DM's choices, from it a priority vector is extracted and the matrix Consistency Ratio is computed which could lead to a correction on the matrix values.

5.3 Final Remarks

Although AHP has been successfully used in a consistent way for many companies, institutions and researches this MCDA methodology has been the target of several critics. During the decade of the 1990's an active discussion between the method's critics and proponents flared up in some major Operational Research and Management journals. The critics aimed, not only, but mostly at two issues: the 19 classification scale proposed by Saaty; and the method's apparent disregard for some of the axioms usually considered in MCD Theory. In the beginning of the XXI century the scientific consensus settled that, in spite of its flaws, AHP is a very useful and pragmatic methodology for solving Multicriteria problems especially when there is a lack of quantitative data, or when the problem involves finding a compromise solution between conflicting Decision Makers (Wallenius et al. 2008).

5.4 Exercises

1. A company needs to hire two new Secretaries. The Human Resources assigns a team of three people to manage the recruiting process. This team is constituted by three people: Amanda, Belmiro and Costa and its task is to interview the candidates and to propose a ranking based on their interview performance, their bachelor global grade and past experience.

Five candidates passed through all recruitment stages with the following results:

	Bachelor Degree Global Grade	Years in a Similar Position	Interview		
			Amanda	Belmiro	Costa
A	14	6	Very Good	Good	Average
B	18	0	Bad	Average	Very Good
C	13	3	Average	Bad	Average
D	15	4	Good	Average	Very Good
E	16	5	Very Bad	Average	Bad

The grade is presented in a 0 to 20 scale.

"Well, for this position I would prefer to hire someone with more experience." – Declared Costa – "For me that is the most important issue. The candidates' behavior during the interview is a close second."

- "Uhm… I am sorry, my friend. But I disagree. For me, an academic good performance is what matters most. We want to keep an image that we only hire the best!" – said Amanda. "But we can agree that the interview's result should be just slightly less important.
- "No way!!!" – Exclaimed Belmiro – "The interview should be much more relevant. It is during the interview that I see if a candidate has the right profile."
- "I do not give a dime, Belmiro!" – Interjected Amanda – "I am the senior here! My opinion is the most relevant."
- "Take it easy, my old friend." – Belmiro answered – "Costa is the one with the least seniority in the company of the three but you are not that much more senior than me! You have to take in consideration my points of view."
- "What a mess…" – Commented Costa – "We could not even agree during the interviews. We can not agree what should be more important for the decision… What should we do? We are going to take ages to rank the candidates."
- "Wait! I recall learning about AHP on my college days! Let me grab a book." – said Belmiro.

Using AHP propose a ranking for the five candidates. The ranking must have an acceptable inconsistency level. For every Judgement Matrix you should assess its inconsistency and correct the matrix if the inconsistency is too high.

In the following questions assume the role of DM Amanda

(a) Build a Judgment Matrix that compares the relative importance of the three criteria from this DM perspective.

(b) For the DM Amanda, consider the criterion "Interview". Build the Judgement Matrix that compares the five candidates.

(c) Repeat the process asked in b) for Criteria "Experience" and "Final Grade".

(d) Obtain the priority vector that ranks the five candidates according to Amanda, integrating the contribution of the three criteria.

In the following questions assume the role of DM Costa

(e) Build a Judgment Matrix that compares the relative importance of the three criteria from this DM perspective.

(f) For the DM Costa, consider the criterion "Interview". Build the Judgement Matrix that compares the five candidates.

(g) Assume that for this DM the matrices proposed in c) also reflect his points of view. Obtain the priority vector that ranks the five candidates according to Costa, integrating the contribution of the three criteria.

Finally, assume the role of the DM Belmiro.

(h) Repeat the processes from questions e) to g) from the DM Belmiro's perspective.

Now, the objective is to integrate all information into a single priority vector that includes all DM's opinions:

(i) Build a Judgement Matrix that compares the relative importance of the three DM, according to their seniority.

(j) Obtain the final priority vector that ranks the five candidates incorporating the opinions of the three DM.

2. AHP Methodology was used in solving a Multicriteria Decision problem with four alternatives (A, B, C e D) and three criteria (1, 2 e 3). The table below presents the priority vectors for the four alternatives from Criterion 1 and Criterion 2 perspectives.

	A	B	C	D
Crit 1	0,18	0,22	0,31	0,29
Crit 2	0,03	0,28	0,23	0,46

For the third Criterion, it is known that A is better than B, and slightly better than C. Alternative D is better than A, absolutely better than B and much better than C. Alternatives C and B are either similar or one is just slightly better than the other.

$$\begin{array}{c} \quad\ 1 \quad\ 2 \quad 3 \\ \begin{array}{c} 1 \\ 2 \\ 3 \end{array}\!\!\left[\begin{array}{ccc} 1 & 3 & 9 \\ 1/3 & 1 & 3 \\ 1/9 & 1/3 & 1 \end{array}\right] \end{array}$$

The Judgment Matrix that compares the relative importance of the three criterion is:

The Random Index for matrices of orders 3 e 4 is, respectively, 0.58 and 0.9.

(a) Comment: "The matrix presented above is transitive."
(b) Built an acceptable Judgment Matrix that compares the Alternatives   form the third criterion's perspective.
(c) What Alternative would you recommend? Justify.

3. Warick, the warrior, and his sorcerous mate, Al-Amyra of Yis, intend to hire a healer to their group of adventurers. They choose the AHP methodology for sorting the four candidates applying for the healer job. Warick and Al-Myra are comparing the candidates in three different criteria: "Experience", "Fumble Rate" and "Temper".

Warick considers that the Criterion "Experience" is more important than "Temper" but that "Temper" is just slightly more important than "Fumble rate".

For Warick the candidates' priority vectors according to the three criteria are:

$$v_{Experience} = [0.71 \ 0.02 \ 0.18 \ 0.09] \quad v_{Fumble \ Rate} = [0.23 \ 0.18 \ 0.03 \ 0.56]$$

$$v_{Temper} = [0.42 \ 0.10 \ 0.12 \ 0.36]$$

For Al-myra the candidates' final priority vector that already incorporates the vectors from the three criteria is

$$v_{Al-myra} = [0.12 \ 0.36 \ 0.34 \ 0.18]$$

(a) Build an acceptable Judgement Matrix that compares the relative importance of the three criteria according to Warick.
(b) Calculate Warick's priority vector for the four candidates.
(c) Knowing that Warick's opinion has more weight than Al-myras's in the final decision that finds the priority vector for the four candidates that incorporate both adventures' points of view.

4. Consider a Multicriteria Decision problem with 4 Alternatives and 3 Criteria.

Consider the following Judgment Matrix that compares all Alternatives according to Criterion 1:

$$A_1 = \begin{bmatrix} 1 & 3 & 5 & 9 \\ 1/3 & 1 & 3 & 1 \\ 1/5 & 1/3 & 1 & 3 \\ 1/9 & 1 & 1/3 & 1 \end{bmatrix}$$

(a) Find the priority vector corresponding to Matrix A_1
(b) Calculate the Consistency Quotient for A_1. Recall that the Random Index for matrix of order 4 is 0.9. Comment your result.
(c) Consider the following Judgement Matrix that compares the importance of

$$A_1 = \begin{bmatrix} 1 & 3 & 1/5 \\ 1/3 & 1 & 1/7 \\ 5 & 7 & 1 \end{bmatrix},$$

and the following priority vectors for the other two criteria:

$$v_{c2} = [0.12 \quad 0.28 \quad 0.39 \quad 0.21] \quad v_{c3} = [0.02 \quad 0.32 \quad 0.21 \quad 0.45]$$

What Alternative would you recommend? Explain every step of your answer.

5. Auntie Leonarda wants to throw a tremendous party to celebrate the beginning of the Social Season. Zumélia, her niece and assistant, suggested some possible themes for the party: the **Aqua Party** (a traditional celebration by the artificial lake in Auntie's mansion grounds, lots of fountains and waterfalls); the **Black Party** (everyone dressed in elegant black clothes, dark decoration and black food and beverages); the **China Party** (Far East themed decoration, Oriental inspired music and clothes, Chinese food); the **Divine Party** (guests must dress like Greek or Egyptian Mythology characters, ethereal and grandiose decoration); the trendy **Electric Party** (neon lights, phosphoric colors, industrial music); and the **Fantasy Party** (dreamy music, surreal and fairy decoration, mazes and labyrinths…)

"To help you out I organized all the information in this table. It considers my assessment of the six alternatives from three criteria's perspective: party cost (in monetary units), impact in the jet set society (in impact points), and organization difficulty (from "Very Easy" to "Very Complex")" – mentioned Zumélia.

Party	Cost	Jet Set Impact	Difficulty
Aqua	2,000	60	Complex
Black	1,200	35	Very Easy
Chinese	500	40	Easy
Divine	3,000	120	Average
Electric	2,500	100	Very Complex
Fantasy	1,500	35	Easy

- "Oh, dear" – Complained Auntie – "You take the fun out of organizing parties. I do not recall worrying about the costs before! This is such a drag."
- "That was in the b.z. Age, Before Zumélia!" – Protested the young assistant – "You hired me to keep your expenses in check! You do not want to go bankrupt, do you? Your parties have been a money drain."
- "All right, all right, But I still think money should not be THAT important. Come now, explain to me what all those numbers mean." – said Auntie Leonarda resignedly.
- "Very well. I estimated the Jet Set impact as the number of celebrities who probably will attend the party. The more, the merrier." Explained Zumélia, "For me, this is absurdly irrelevant, but because I know you take this kind of stuff very seriously, you did the estimations based on your historical data on party attendance."

"Oh, yes, my dear! That is absolutely primordial!", agreed Auntie. "I am just nor sure, I want more than 100 pseudo-VIPs in the party. I will not be able to give everyone attention. Now, about the organization difficulty. Why is that an issue? Yes, a more complex party can bring some headaches but it is not like it is going to be the End of the World."

- "Yeah, right. But I am the one who is going to have to take care of everything so I beg to differ", said Zumélia with a pinch of sarcasm. "And let me add that for me this criterion is almost as important as the party cost."
- "I doubt we will ever agree. Why can't we just have the party I want?" complained Auntie.

"Because, my dear Aunt, your family wants me to reign your superfluous expenses. You agreed to hire me or they will try to declare you incompetent and sent you to a rest home, Bear with me. We can get along and make this a fun thing."

Use AHP to decide what party should occur. The analysis has to include both Zumélia's and Auntie Leonarda's points of view.

6. The Government of Lusoland is considering the viability of building a third bridge across river Tajus in Lisbonia metropolitan area. There are four possible places where the bridge can be built. The four alternatives were evaluated according to three points of view: Cost (in millions of Eurons); Environmental Impact (in a scale of 0 - Null to 20 - catastrophic); and Road Traffic relief (in a qualitative scale "Null, Very weak, Weak, Average, High, Very High, Extreme).

The information concerning the evaluations is presented below:

Possible crossing	Cost	Environmental Impact	Traffic Relief
Arreiro – Lisbonia	300	9	Average
Brandão – Lisbonia	150	6	Average
Ceixal – Lisbonia	100	5	Weak
Drafaria – Lisbonia	500	18	Extreme

Zeca Plato, the Prime Minister (PM), meets with three of his Ministers: the talkative and annoying Public Works Minister (PWM), the aggressive and bludgeoning Environmental Minister (EM) and the cold and rational Finance Minister (FM).

"Ok, Ladies and Gentleman. We have to announce a decision today or I will be eaten alive next time I go to the Parliament." – informed the PM.

"As we all know, the Cost is obviously much more important than the other criteria. I have with me the financial analysis and I can tell that, according to our budget: a cost of 100 is much more interesting than a cost of 500 and more interesting than a cost of 300. On the other hand, a cost of 300 is just slightly more interesting than a cost of 500", declared the FM.

"Well, my dear, I believe that we all agree with you in general, BUT" , started the EM – That Economic myopic vision has been crushing our beautiful country's rich biodiversity! Our Government MUST start putting our Environmental agenda ahead of all other concerns! I recommend that the Environmental Impact should be extremely more important than the other two!"

"Not in a million years, Madam!" – Vociferated the PWM – "The bridge will be built to improve our taxpayers' Life. The criteria "Traffic relief" should be much more important than the Cost and plainly more important than the Environmental Impact. And I have to add that for me an Environmental Impact of 5 or 6 are exactly the same, In my not so humble opinion the relevant point here is that an Extreme effect in "traffic relief" is much more interesting than an average effect, and extremely more interesting than a weak effect, pardon me the pun, please."

"Very, well, Let us all assume that I will adopt the specialist's opinion. FM's for the cost, PWM's for the traffic relief and EM's for the Environmental Impact. Just let me digest what you three said, OK? Nevertheless, let me ask you all an additional question: Everyone agrees on the rest?" , Asked the PM.

"I am very sorry, but I cannot agree with what the PM said. For me, an Average Traffic Relief has to be much more interesting than a weak one", answered the FM, "If I am giving money for a new bridge it better show me some noticeable effect on the freaking traffic! And especially for our Green Environmental friend here, remember that less traffic during the rush hours means less CO_2 emissions." Finished the FM with a calculated smile on her lips.

— "Oh, come on. You are just like kids. Always arguing." Stated the PM calmly
"Well… I am in charge here. Let us do this using AHP. And my opinion is the Law around here."

Use AHP to help the Government of Lusoland to decide where to build the bridge. The opinions of all Ministers should be incorporated in the final priority vector.

References

Bebiano, N., R. Fernandes and S. Furtado (2020). Reciprocal matrices: Properties and approximation by a transitive matrix. *Computational and Applied Mathematics*, 39(2), 1–22. https://doi.org/10.1007/s40314-020-1075-2

Horn, R. and C. Johnson (1985). *Matrix Analysis*. Cambridge University Press.

Saaty, T.L. (1977). A scaling method for priorities in hierarchical structures. *Journal of Mathematical Psychology*, 15(3), 234–281. https://doi.org/10.1016/0022-2496(77)90033-5

Saaty, T. (1980). *The Analytical Hierarchy Process*. McGraw-Hill.

Shiraishi, S., T. Obata and M. Daigo (1998). Properties of a positive reciprocal matrix and their application to AHP. *Journal of the Operations Research Society of Japan*, 41(3), 413–414. https://doi.org/10.15807/jorsj.41.404

Wallenius, J., J.S. Dyer, P.C. Fishburn, R.E. Steuer, S. Zionts and K. Deb (2008). Multiple criteria decision making, multiattribute utility theory: Recent accomplishments and what lies ahead. *Management Science*, 54(7), 1336–1349. https://doi.org/10.1287/mnsc.1070.0838

TOPSIS and Fuzzy TOPSIS

6.1   Introduction

TOPSIS, Technique for Order Preference by Similarity to Ideal Solution, was proposed by Hwang and Yoon in 1981, as a method to solve Multi Criteria Decision (MCD) problems. This technique is based on the assumption that the Decision Maker (DM) wishes to find the alternative that is closest to the ideal solution and the farthest from the nadir solution (or the anti-ideal solution or the negative-ideal solution as it has been named by the authors). The ideal solution is a non-existing alternative that is composed of the best values within each of the criteria in evaluation. It uses weights to model the DM preferences concerning the criterias relative importance. One thing that makes TOPSIS a very attractive method is its simplicity, not only conceptually but also regarding computation issues. Behzadian et al. (2012) reviewed 266 works that were published in more than 100 journals since the year 2000. Moreover, the method's versatility has been explored and several alternatives and modifications concerning the modelling of weights, the definition of the ideal and nadir solutions and the distance measuring to these solutions have been proposed over the years (Papathanasiou and Ploskas 2018).

In many real-world problems, several sources of uncertainty may exist when evaluating alternatives with respect to criteria or when assessing weights. One of the easiest ways from the DM to make these evaluations (or to provide information concerning criteria relative importance) is by using natural language. A fuzzy set approach is quite common in MCD the literature "translate" natural language into numerical data. Since the TOPSIS method has been extended to the fuzzy environment and it has been very well accepted by academia, the first version of fuzzy TOPSIS will also be presented (Palczewski and Salabun 2019, Salih et al. 2019).

This chapter is organized as follows. It starts with the presentation of the TOPSIS method with its six-step approach. During its presentation, the example that has been solved in the previous chapters (and described in Appendix) will be solved allowing the reader to better perceive the differences among MCD methods. Then, section 6.3 addresses some of the method's limitations. The Fuzzy

TOPSIS method using triangular fuzzy numbers is presented in section 6.4. Since the concept of fuzzy numbers is not common in most academic curricula, a short introduction to fuzzy sets and fuzzy numbers provides the main concepts needed to understand the method descriptions (sections 6.4.1 and 6.4.2, respectively). How to integrate a group of decision makers in Fuzzy TOPSIS is presented in section 6.4.3. The chapter ends with some final remarks and a set of exercises covering all the topics discussed.

6.2   The Method

The TOPSIS method starts, as any other MCD methods, with the identification of the alternatives and criteria. Then it proceeds with the assessment of the alternatives' performance within the different criteria (Ishizaka and Nemery 2013). Let's suppose one has a set of n alternatives evaluated by k criteria that have been organized in a Decision Matrix with n rows and k columns. Table 6.1 presents a generic decision matrix where A_i, $i = 1, ..., n$ is the ith alternative, C_j, $j = 1, ..., k$ is the jth criterion and x_{ij} is the performance of alternative A_i with respect to criterion C_j.

Table 6.1: Generic Decision Matrix

	C_1	C_2		C_j		C_k
A_1	x_{11}	x_{12}	...	x_{1i}	...	x_{1k}
A_2	x_{21}	x_{22}	...	x_{2j}	...	x_{2k}
$\vdots$	$\vdots$	$\vdots$		$\vdots$		$\vdots$
A_i	x_{i1}	x_{i2}	...	x_{ij}	...	x_{ik}
$\vdots$	$\vdots$	$\vdots$		$\vdots$		$\vdots$
A_n	x_{n1}	x_{n2}	...	x_{nj}	...	x_{nk}

Criteria may be increasing or decreasing in scale. This is to mean that given two alternatives A_i and A'_i, for the increasing scale criterion C_j (a "benefit" criterion), one has $x_{ij} < x'_{ij}$ then, for the DM, alternative A'_i is preferable to alternative A_i. On the contrary, if criterion C_j is "cost" criterion (decreasing scale criterion) then when $x_{ij} < x'_{ij}$, the DM considers alternative A_i to be preferable to alternative A'_i. If some criterion is defined in a qualitative scale, prior to the method's application, the DM should be questioned so as to convert it into a quantitative scale. The authors do not propose any specific method or approach, therefore, one can use any method available in the literature, as for instance, the methods addressed in SMART methodology (chapter 3).

Table 6.2 summarizes the data from the example presented in the Appendix where four alternatives have been evaluated according to three perspectives. For the Team Leader of the research group, the attributes "Interview" and "Theoretical test" are presented in increasing scales ("the higher the value, the better" and, therefore, they are "benefit" criteria), while attribute "Demerits in practical Test" is in a decreasing scale ("the smaller the value, the better" and thus, it is a "cost" criterion).

Table 6.2: Example's Data

		Interview	Demerits in Practical Test	Theoretical Test
	A	Very Good	4	75.4%
	B	Good	2	84.6%
Candidates	C	Regular	3	95.7%
	D	Good	6	90.2%
	E	Very Good	6	70.5%

The "Interview" has been evaluated using a qualitative scale, it then needs to be "converted" into a numerical attribute. The Direct Rating proposed in SMART is one possible method to perform this transformation. Let's then make use of the values calculated in chapter 3. Table 6.3 is the Decision Matrix which is the starting point for TOPSIS.

Table 6.3: Example's Decision Matrix

	C_1	C_2	C_3
A	100	4	0.754
B	60	2	0.846
C	30	3	0.957
D	60	6	0.902
E	100	6	0.705

The TOPSIS method follows a set of steps that will be detailed below. Since no distinction is made concerning the terms attribute and criterion, they will be used interchangeably throughout this chapter.

Step 1: Normalize the Decision Matrix

This step removes dimensionality from the criteria, allowing them to be comparable. There are several methods to render criteria non–dimensional. However, the authors suggested the use of the Euclidean norm. Let r_{ij} be the element of the normalized Decision Matrix, R, corresponding to the pair alternative i, criterion j:

$$r_{ij} = \frac{x_{ij}}{\sqrt{\sum_{i=1}^{n} x_{ij}^2}}$$

This step renders each criterion with the same unit length vector.

Let's take alternative A and calculate the corresponding normalized values:

$$r_{A1} = \frac{100}{\sqrt{100^2 + 60^2 + 30^2 + 60^2 + 100^2}} = 0.5965, \qquad r_{A2} = \frac{4}{10.0499} = 0.398$$

and $r_{A3} = \dfrac{0.754}{1.8737} = 0.4024$.

These values will be the first of the normalized decision matrix (Table 6.4). Notice, for ease of readability, all values have been rounded to the 4th decimal place[1].

Table 6.4: Normalized Decision Matrix

	C_1	C_2	C_3
A	0.5966	0.398	0.4024
B	0.3579	0.199	0.4515
C	0.179	0.2985	0.5108
D	0.3579	0.597	0.4814
E	0.5966	0.597	0.3763

Step 2: Compute the Weighted Normalized Decision Matrix

Since the different criteria may have different importance to the Decision Maker, the TOPSIS method incorporates this information through weights. Again, the method's authors do not explicitly propose a strategy to extract this information from the DM. Therefore, one can apply a method that has been proposed in the literature.

Let $W = (w_1, w_2, ... w_j ..., w_k)$, $\sum_{j=1}^{k} w_j = 1$ be a weight vector that reflects the DM's relative importance given to each attribute. Let v_{ij} be the element of the weighted normalized decision matrix, V, corresponding to the pair alternative i, criterion j:

$$v_{ij} = w_j \cdot r_{ij}$$

Suppose the DM wishes the weights to be $w_1 = 0.5$, $w_2 = 0.4$, and $w_3 = 0.1$ for the attributes interview, demerits and theoretical test, respectively. Taking an alternative A, let's calculate the corresponding weighted values:

$$v_{A1} = 0.5 \cdot 0.5965 = 0.2983, \ v_{A2} = 0.4 \cdot 0.398 = 0.1592 \text{ and}$$
$$v_{A3} = 0.1 \cdot 0.4024 = 0.04024.$$

Step 3: Determine the Ideal and Nadir Solutions

The next step is the extraction of the ideal and nadir solutions from the weighted

---

[1] A spreadsheet implementation is one of the easiest ways to perform all of the method's calculations, since its steps are friendly with a spreadsheet environment. If a spreadsheet is used, number rounding should not be done to avoid error propagation.

Table 6.5: Weighted Normalized Decision Matrix.

	C_1	C_2	C_3
A	0.2983	0.1592	0.0402
B	0.179	0.0796	0.0452
C	0.0895	0.1194	0.0511
D	0.179	0.2388	0.0481
E	0.2983	0.2388	0.0376

normalized decision matrix. These will act as two fictitious alternatives to which the distance of the real alternatives will be measured against. As mentioned above, these solutions are closely related with each type of criteria. If a criterion represents a "benefit" then the best value will be the highest of the corresponding column in the weighted normalized decision matrix. If it represents a "cost", then the best value will be the lowest of the same column.

Let A^* and A^- be the ideal and nadir solutions. These are defined as

$$A^* = \{(v_{ij} : j \in J), (v_{ij} : j \in J') : i = 1, ..., n\}$$
$$= (v_1^*, v_2^*, ..., v_j^*, ..., v_k^*), \text{ and}$$
$$A^- = \{(v_{ij} : j \in J), (v_{ij} : j \in J') : i = 1, ..., n\}$$
$$= (v_1^-, v_2^-, ..., v_j^-, ..., v_k^-),$$

where $J = \{j = 1, ..., k: j$ is a criterion associated with benefits$\}$ and $J' = \{j = 1, ..., k: j$ is a criterion associated with costs$\}$.

In the example, the criteria associated with benefits are "Interview" (C_1) and "Theoretical test" (C_3) and the criterion "Demerits" (C_2) is associated with costs. So, the ideal and the nadir solutions are respectively:

$$A^* = \{(v_{ij} : j \in \{C_1, C_3\}), (v_{ij} : j \in \{C_2\}) : i = 1, ..., n\}$$
$$= (0.2983, 0.0796, 0.0511), \text{ and}$$
$$A^- = \{(v_{ij} : j \in \{C_1, C_3\}), (v_{ij} : j \in \{C_2\}) : i = 1, ..., n\}$$
$$= (0.0895, 0.2388, 0.0376)$$

Step 4: Calculate the Distances to the Ideal and Nadir Solutions

In this step, one needs to calculate the distance of each alternative from the ideal and the nadir solutions. The authors proposed the use of the Euclidean distance[2]. Thus, the distance of each alternative to the ideal solution is given by

$$S_i^* = \sqrt{\sum_{j=1}^{k} (v_{ij} - v_j^*)^2}, i = 1, ..., n$$

---

[2] In chapter 9, when addressing distance minimization multi–objective models further insights are provided with respect to distance functions.

while the distance to the nadir solution is given by

$$S_i^- = \sqrt{\sum_{j=1}^{k} (v_{ij} - v_j^-)^2}, \; i = 1, ..., n.$$

Taking alternative A, these distances are

$$S_A^* = \sqrt{\sum_{j=1}^{3} (v_{Aj} - v_j^*)^2}$$

$$= \sqrt{(0.2983 - 0.2983)^2 + (0.1592 - 0.0796)^2 + (0.04024 - 0.05108)^2}$$

$$= 0.0803$$

and

$$S_A^- = \sqrt{\sum_{j=1}^{3} (v_{Aj} - v_j^-)^2}$$

$$= \sqrt{(0.2983 - 0.0895)^2 + (0.1592 - 0.2388)^2 + (0.04024 - 0.05108)^2}$$

$$= 0.2235.$$

Table 6.6 shows the Euclidean distance to the ideal and nadir solutions for the five alternatives in evaluation. One can see that alternative A is the closest to the ideal solution and is also the one further away from the nadir solution. Therefore, it appears to be a very good candidate to be the best alternative, meaning, that the one the DM should opt for. However, it is frequent that the best alternative with respect to one of the solutions does not perform well with respect to the other. For instance, alternative C is second best concerning the distance to the ideal solution but is the one closer to the nadir point. Moreover, if one wishes/needs to perform a full ranking, these two measures need to be aggregated.

Table 6.6: Euclidean Distances to the Ideal (S^+) and Nadir (S^-) Solutions

	S^+	S^-
A	0.0803	0.2235
B	0.1194	0.1828
C	0.2126	0.1202
D	0.199	0.0901
E	0.1598	0.2088

Step 5: Calculate the Relative Closeness to the Ideal Solution

In order to aggregate the two previous distances, the authors propose the relative closeness to the ideal solution as the index

$$C_i^* = \frac{S_i^-}{S_i^+ + S_i^-}, i = 1, ..., n.$$

As defined, $0 < C_i^* < 1$, $i = 1$,..., n. Notice that, when $C_i^* = 1$, then $i = A^*$ and when $C_i^* = 0$, then $i = A^-$. So, the highest the value of C_i^*, the closer the alternative is to the ideal solution.

Table 6.7 shows the relative closeness to the ideal solution for all five alternatives.

Table 6.7: Relative Closeness to the Ideal Solution (C^*)

	C^*
A	0.7357
B	0.6049
C	0.3612
D	0.3117
E	0.5665

Step 6: Final Ranking

The method's last step is to sort the alternatives from the most preferred one to the least one. As defined, the higher the value of the relative closeness to the ideal solution, the better the solution. Thus, the alternatives can be preference ranked according to the descending other of C_i^*.

In the example, candidate A is the one that performs better according to the modelled DM's preferences, since it is the one with the highest score (relative closeness). Table 6.8 shows the final rank, which is

$$1^{st} - A; 2^{nd} - B; 3^{rd} - E; 4^{th} - C; 5^{th} - D.$$

Table 6.8: Alternatives Ranking

	C^*	rank
A	0.7357	1
B	0.6049	2
E	0.5665	3
C	0.3612	4
D	0.3117	5

6.3 Some of the Method's Limitations

As any other method that has been proposed for MCD making, TOPSIS limitations have been extensively addressed. Among the most relevant are the normalization technique, weights and rank reversal. These are not exclusive to TOPSIS (e.g. the SMART methodology has been criticized because the use of weights and rank reversal is one of the limitations pointed out to AHP). In this section the three above-mentioned drawbacks will be looked at in some detail and suggestions on how to overcome them will be given.

6.3.1 Normalization Step

Concerning the impact the normalization method has on the final ranking, there are two aspects that call for attention. One is the unit in which the criterion is defined and the other concerns the normalization technique in itself. A temperature attribute is an example of the former. It might be measured in Celsius or in Fahrenheit, and one wishes the same final ranking irrespective of the unit chosen. However, Opricovic and Tzeng (2004) showed that this might not be true. Concerning the latter, Vafaei et al. (2018) performed a comprehensive study on the impact the TOPSIS normalization step has on final ranking. The authors compared six different normalization techniques: Euclidean normalization, three linear normalizations (Max, MaxMin and Sum), logarithmic normalization, and fuzzy normalization. The last mentioned normalization technique was applied in its simplest (for further details refer to Ribeiro 1996). To assess the most appropriate normalization technique to be applied in TOPSIS, the authors proposed a three-step approach so as to have a consistent evaluation of the results. An illustrative example is presented as a small tutorial on how to apply the proposed approach. From this example, the authors were able to conclude that the Euclidean normalization appears the most suitable for TOPSIS, while the logarithmic normalization performed the worst in the evaluation process. For the remaining tested techniques, two of the linear normalizations ranked second and third (Max and Sum, respectively). The MaxMin and the fuzzy approaches provided inconclusive results.

Let's apply the three "best" normalization techniques to the example and analyze their impact on the final ranking. In the Max normalization technique, all values of each criterion are divided by the largest one observed amongst them. Two different expressions are presented depending on the type of criterion (benefit or cost criterion). Let r_{ij} be the element of the normalized decision matrix, R, corresponding to the pair alternative i, criterion j, then

$$r_{ij} = \frac{x_{ij}}{x_j^+}, i = 1, ..., n \text{ and } j \in J,$$

$$r_{ij} = 1 - \frac{x_{ij}}{x_j^+}, i = 1, ..., n \text{ and } j \in J',$$

where $x_j^+ = \max\limits_{i=1,...,n} x_{ij}$, and J and J' as defined in step 3.

The Sum normalization technique also differentiates benefit from cost criteria, proposing one expression for each criterion type:

$$r_{ij} = \frac{x_{ij}}{\sum_{i=1}^{n} x_{ij}}, i = 1, ..., n \text{ and } j \in J,$$

$$r_{ij} = \frac{1/x_{ij}}{\sum_{i=1}^{n} 1/x_{ij}}, i = 1, ..., n \text{ and } j \in J'.$$

Note that, after performing any of these normalization techniques and since the cost and benefit criteria were distinguished with a different expression, the normalized values now have the same "orientation". Meaning that the higher the value the better the alternative. Consequently, when determining the ideal (nadir) solution, at step 4, the maximum (minimum) value of each criterion should be chosen for the ideal (nadir) solution. Table 6.9 and Table 6.10 show the normalized values when applying the Max and the Sum techniques, respectively.

Table 6.9: Max Normalization Decision Matrix

	C_1	C_2	C_3
A	1	0.3333	0.7879
B	0.6	0.6667	0.884
C	0.3	0.5	1
D	0.6	0	0.9425
E	1	0	0.7367

Table 6.10: Sum Normalization Decision Matrix

	C_1	C_2	C_3
A	0.2857	0.1765	0.1811
B	0.1714	0.3529	0.2032
C	0.0857	0.2353	0.2298
D	0.1714	0.1176	0.2166
E	0.2857	0.1176	0.1693

Assuming the same weight vector and applying the remaining steps, one reaches quite similar rankings (Table 6.11 and Fig. 6.1). Only the two first alternatives change position when applying the Sum normalization technique. This allows the conclusion that the two best alternatives are in fact A and B, with A being ranked first by two methods. If the Team Leader is still not sure about which candidate to choose, the procedure proposed by Vafaei et al. (2018) can be applied.

Table 6.11: Relative Closeness to the Ideal Solution and Final Ranking for
Three Normalization Techniques

	Euclidean		Max		Sum	
	C^*	Rank	C^*	Rank	C^*	Rank
A	0.7357	1	0.7349	1	0.5922	2
B	0.6049	2	0.6046	2	0.6437	1
C	0.3612	4	0.3615	4	0.3006	4
D	0.3117	5	0.3123	5	0.2811	5
E	0.5665	3	0.5663	3	0.5144	3

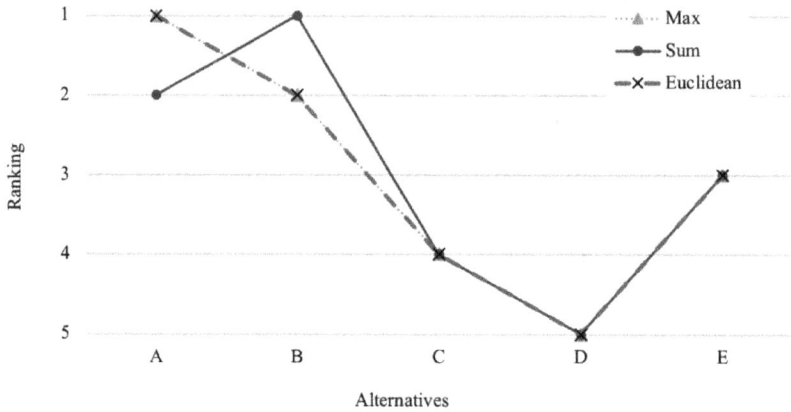

Figure 6.1: Alternatives Ranking for Three Normalization Techniques

6.3.2   Weights

One of the pieces of criticism concerning weights is related to the lack of an elicitation method[3]. This issue is, however, easily overcome since several techniques have been proposed that can be applied when the weights need to be known (Huang and Li 2012). The Direct Rating proposed in SMART or the AHP approach are methods addressed in the previous chapter that can be used to compute weights[4]. Irrespective of the method's choice, a sensitivity analysis can be performed to investigate the impact small changes in one of the weights may have in the final ranking.

The example final ranking is: 1st – A; 2nd – B; 3rd – E; 4th – C; 5th – D. One might wonder how robust this ranking is. Or, in other words, if a small change is made on the weight of the most important criterion, what would be

---

[3] A method to gather information from the Decision Maker.
[4] Olsen (2004) compared several methods to assess weights.

the impact on the final ranking. Let's then vary the weight value of criterion C_1, while keeping unchanged the weight relation between the remaining criteria. Remember, the weights initially proposed by the DM are $w_1 = 0.5$, $w_2 = 0.4$, and $w_3 = 0.1$. Figure 6.2 shows the relative closeness index (C^*) for all alternatives when considering different weight values for criterion C_1. Since the higher the C^* the better the alternative is ranked, the figure also shows the ordering of the different alternatives, according to the different weights. It can easily be seen that closer to the weight value suggested by the DM, the alternatives placed first and second change positions. This means the DM needs to be pretty sure about the weight given to criterion C_1, since if it falls below 0.5, then alternative B will be a very good candidate to be the best choice. This additional procedure with very simple computation may provide valuable information to support the DM choice.

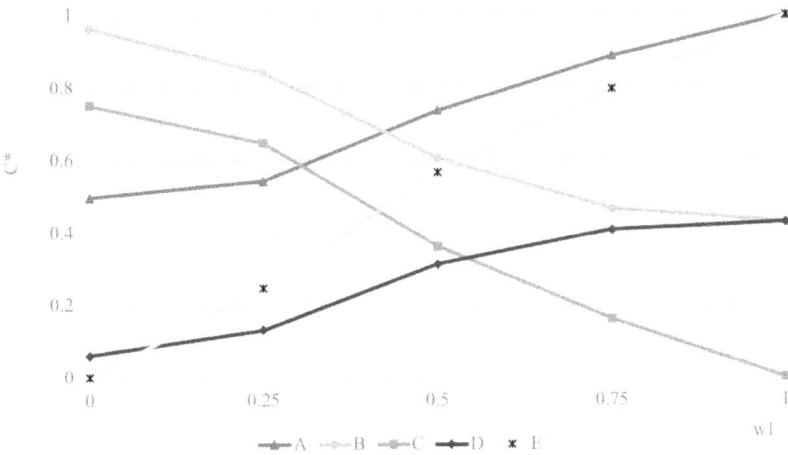

Figure 6.2: Relative Closeness (C^*) for the Five Alternatives when Considering Different Weight Values for Criterion C_1 (Sensitivity Analysis for w_1)

6.3.3 Rank Reversal

Rank reversal is a phenomenon that occurs in many MCD making problems where, when adding a new alternative or removing an existing one, the final ranking changes considerably. Cases happen where the final ranking is reversed when an alternative has been added/removed (Papathanasiou and Ploskas 2018). This phenomenon is not exclusive to TOPSIS. It is in fact common to many MCD problems (Wang and Luo 2009). In TOPSIS this limitation has two origins: the use of the Euclidean norm at the normalization step and the choice of the ideal and nadir solutions. García-Cascales and Lamata (2012) and de Farias Aires and Ferreira (2019) study strategies that minimize, if not overcome, this limitation. The former focus on the impact of removing one alternative and point out the causes that lead to rank reversal. The latter extends this work and proposes small changes on the original TOPSIS method rendering it "immune" to rank reversal.

As changes the authors suggest the use 1) of a new parameter named "domain" representing the value range each criterion can take; 2) the Max or MaxMin as normalization techniques, where the max and min values concern the domain values and not the alternatives performances on the criteria, and 3) the ideal and nadir solutions are fixed (they are the domains extreme values, which remain unchanged when adding or removing alternatives). This work also presents other reasons that cause rank reversal in several MCD methods.

6.4   Fuzzy TOPSIS

The use of qualitative evaluations is very often much easier for the DM than to provide accurate values (known in fuzzy terminology as *crisp values*). However, expressions as "very good", "insufficient", "very important" do not have a clear and definite meaning. Zadeh (1975) defined **linguist variables** as variables "whose values are words or sentences in a natural or artificial language. For example, *Age* is a linguistic variable if its values are linguistic rather than numerical, i.e., *young, not young, very young, quite young, old, not very old* and *not very young*, etc., rather than 20, 21, 22, 23...". These variables appear, in other works, with different designations. For instance, in SMART methodology (chapter 3) they are known as qualitative scales and the Direct Rating method was presented as a way to translate this subjective judgment into quantitative value. Other methods may be applied and one that has been well developed within the MCD context in the use of fuzzy sets. Bellman and Zadeh (1970), in their seminal work on decision making in fuzzy environments, informally defined **fuzzy sets** as "classes of objects in which there is no sharp boundary between those objects that belong to the class and those that do not". This concept has been integrated into several methods and methodologies to solve MCD problems.

  The method described below was proposed by Chen (2000) and assumed both weights and the performance evaluation of alternatives in each criterion as linguist variables. Before entering further into the method, some definitions related to fuzzy concepts are given so as to better understand how the FTOPSIS method works. By no means, does one wishes to present all the mathematical background. Rather, the next section introduces a few definitions of important concepts. For more details and theoretical background Hanss (2005), Wang et al. (2009), Salih et al. (2019), or Yatsalo et al. (2021) are good references.

6.4.1   Fuzzy Sets and Fuzzy Numbers

A **fuzzy set** A in X is a set of ordered pairs

$$A = \{(x, \mu_A(x)), x \in X\}$$

where X in a collection of objects, $\mu_A(x)$ is the **membership function** of $x \in A$, with $\mu_A: X \rightarrow [0, 1]$. A **fuzzy number** (FN) is fuzzy set in X that has two properties: it is normal and convex. The former property is quite important for the Fuzzy TOPSIS method since it assures that at least one element of X has a membership value of 1. Amongst the most common fuzzy numbers are triangular FN and trapezoidal FN. A **triangular FN** is defined by its membership function $\mu_A(x)$, shown in Fig. 6.3. The most common notation for a triangular FN is its triplet $A = (a, b, c)$.

$$\mu_A(x) = \begin{cases} 0, & x < a \vee x > c \\ \dfrac{x-a}{b-a}, & a \le x \le b \\ \dfrac{c-x}{c-b}, & b \le x \le c \end{cases}$$

Although a, b and c may be any real number, the Fuzzy TOPSIS method proposed by Chen (2000) makes use of non-negative triangular FNs. Thus, from this point on, when referring to triangular FNs one is assuming $c \ge b \ge b \ge a \ge 0$. A detailed approach on how to build triangular fuzzy numbers can be found in Chen (2004).

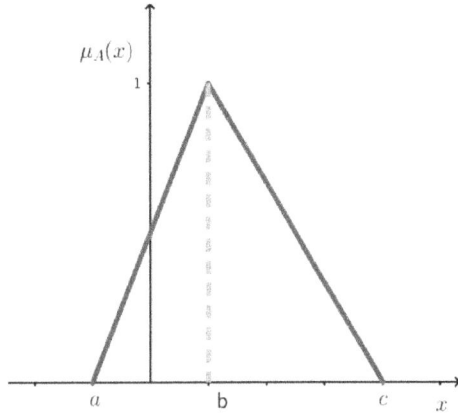

Figure 6.3: Triangular Fuzzy Number

6.4.2 The Method

The FTOPSIS follows similar steps to the original method and it also starts by building the decision matrix. Let $\tilde{D}$ be the Decision Matrix for MCD problem where the performance of each of the n alternatives on the k criteria, is evaluated using linguist variables $\tilde{x}_{ij}$, $i = 1, \dots n$ and $j = 1, \dots, k$ (Table 6.12).

Table 6.12: Generic Fuzzy Decision Matrix

	C_1	C_2		C_3
A_1	$\tilde{x}_{11}$	$\tilde{x}_{12}$	$\cdots$	$\tilde{x}_{1k}$
A_2	$\tilde{x}_{21}$	$\tilde{x}_{22}$	$\cdots$	$\tilde{x}_{2k}$
$\vdots$	$\vdots$	$\vdots$		
A_n	$\tilde{x}_{n1}$	$\tilde{x}_{n2}$	$\cdots$	$\tilde{x}_{n3}$

Let $\tilde{W} = (\tilde{w}_1, \tilde{w}_2, \dots, \tilde{w}_j, \dots, \tilde{w}_k)$ be the importance weight of each criteria also defined by linguist variables. Without loss of generality, let's assume these linguist variables are described by triangular FNs such as

$\tilde{x}_{ij} = (a_{ij}, b_{ij}, c_{ij})$ and $\tilde{w}_j = (w_j^1, w_j^2, w_j^3)$. If other FNs are used, adequate adaptation should be made.

Let's make some changes on the example that has been solved above to account for FNs. Suppose the benefit criteria "Interview" and the classification on the "Theoretical test" have been evaluated by the Team Leader with linguist variables "Very Bad", "Bad", "Regular", "Good", "Very Good". For the criterion "Demerits in practical test" different linguist variables have been used. Table 6.13 shows the DM's evaluation of all alternatives using such variables.

Table 6.13: Example's with Linguistic Data

		Interview (C₁)	Demerits in Practical Test (C₂)	Theoretical Test (C₃)
	A	Very Good	Average	Good
	B	Good	Very Low	Bad
Candidates	C	Regular	Low	Very Good
	D	Good	High	Good
	E	Very Good	High	Regular

The translation of these linguist variables as a triangular FNs are given in Table 6.14 and Table 6.15, and depicted in Fig. 6.4 and Fig. 6.5. This step, the definition of the triangular membership functions for the qualitative scales, is known as "fuzzification".

Table 6.14: Linguist Variables for Benefit Criteria (C_1 and C_3)

Very Bad (VB)	(0,0,2)
Bad (B)	(0,2,4)
Regular (R)	(3,5.5,8)
Good (G)	(6,8,10)
Very Good (VG)	(9,10,10)

Table 6.15: Linguist Variables for the Cost Criterion (C_2)

Very High (VH)	(1,1,2)
High (H)	(1,2,3)
Average (Av)	(2,4,6)
Low (L)	(5, 7, 9)
Very Low (VL)	(8, 9, 10)
Absent (A)	(9, 10, 10)

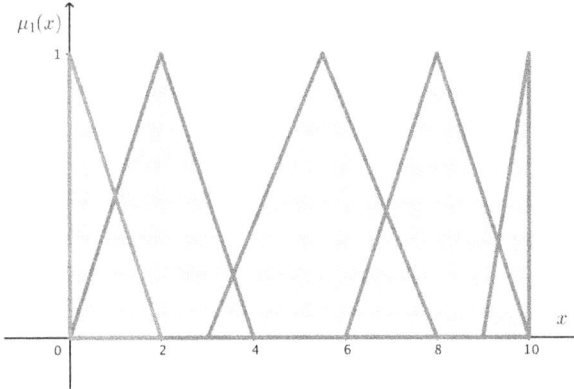

Figure 6.4: Graphical Representation of Benefit Criteria Linguist Variable

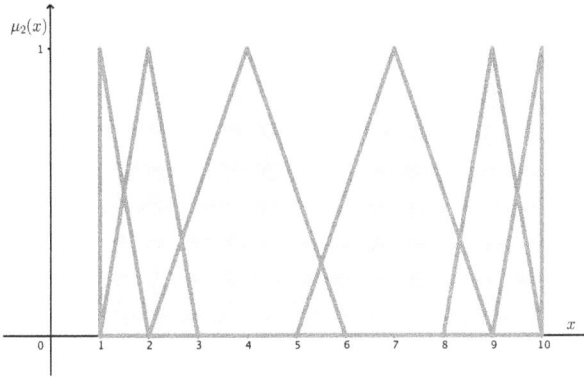

Figure 6.5: Graphical Representation of Cost Criterion Linguist Variable

The decision matrix with linguistic variables, Table 6.13, can now be used to write using the triangular FNs defined in Table 6.16.

Table 6.16: Fuzzy Decision Matrix

	C_1	C_2	C_3
A	(9, 10, 10)	(2, 4, 6)	(6, 8, 10)
B	(6, 8, 10)	(8, 9, 10)	(0, 2, 4)
C	(3, 5.5, 8)	(5, 7, 9)	(9, 10, 10)
D	(6, 8, 10)	(1, 2, 3)	(6, 8, 10)
E	(9, 10, 10)	(1, 2, 3)	(3, 5.5, 8)

Notice that, although criterion C_2 is (originally) a cost criterion, the fuzzification transformed it into a "benefit" criterion since now low values ("low" triangles) correspond to "bad" evaluations, while larger values (triangles placed higher in the x-axis) correspond to alternatives with better evaluations.

Step 1F: Fuzzy Normalization

Since the Euclidean normalization for triangular FNs produces quite complicated functions and given that other normalization techniques can be applied, a linear normalization was proposed by Chen (2000). This linear normalization assures the property that triangular FNs ranges are defined in [0, 1].

Let $\tilde{r}_{ij}$ be the element of the normalized fuzzy decision matrix, $\tilde{R}$, corresponding to the pair alternative i, criterion j such that

$$\tilde{r}_{ij} = \left(\frac{a_{ij}}{c_j^*}, \frac{b_{ij}}{c_j^*}, \frac{c_{ij}}{c_j^*} \right), j \in J$$

$$\tilde{r}_{ij} = \left(\frac{a_j^-}{c_{ij}}, \frac{a_j^-}{b_{ij}}, \frac{a_j^-}{a_{ij}} \right), j \in J'$$

where $c_j^* = \max_{i=1,...,n} c_{ij}, j \in J$ and $a_j^- = \min_{i=1,...,n} a_{ij}, j \in J'$ where J and J' are the set of benefit criteria and cost criteria, respectively (as defined in step 3, section 6.2). For cost criteria, attention should be taken when fuzzifying the linguistic variables since this normalization technique imposes that a_{ij}, b_{ij} and c_{ij}, $i = 1, ..., n, j = 1, ..., k$, cannot be zero.

Taking our example, one has now three benefit criteria. Thus, $C_1^* = (10, 10, 8, 10, 10) = 10$, $C_2^* = (6, 10, 8, 3, 3) = 10$ and $C_3^* = (10, 4, 10, 10, 8) = 10$. The normalized values of alternative A for three criteria are:

$$\tilde{r}_{A1} = \left(\frac{a_{A1}}{c_1^*}, \frac{b_{A1}}{c_1^*}, \frac{c_{A1}}{c_1^*} \right) = (0.9, 1, 1),$$

$$\tilde{r}_{A2} = \left(\frac{a_{A2}}{c_2^*}, \frac{b_{A2}}{c_2^*}, \frac{c_{A2}}{c_2^*} \right) = (0.2, 0.4, 0.6),$$

$$\tilde{r}_{A3} = \left(\frac{a_{A3}}{c_3^*}, \frac{b_{A3}}{c_3^*}, \frac{c_{A3}}{c_2^*} \right) = (0.6, 0.8, 1).$$

Table 6.17 shows the normalized fuzzy decision matrix. Notice that if the same triangular FN had been used when evaluating the three criteria, then C_2 would still be considered a cost criterion. In that case, the second expression for $\tilde{r}_{ij}$ should have been used.

Table 6.17: Fuzzy Normalized Decision Matrix

	C_1	C_2	C_3
A	(0.9, 1, 1)	(0.2, 0.4, 0.6)	(0.6, 0.8, 1)
B	(0.6, 0.8, 1)	(0.8, 0.9, 1)	(0, 0.2, 0.4)
C	(0.3, 0.55, 0.8)	(0.5, 0.7, 0.9)	(0.9, 1, 1)
D	(0.6, 0.8, 1)	(0.1, 0.2, 0.3)	(0.6, 0.8, 1)
E	(0.9, 1, 1)	(0.1, 0.2, 0.3)	(0.3, 0.55, 0.8)

Step 2F: Fuzzy Weighting

The relative importance among criteria can also be evaluated using a qualitative scale (or some linguist variables). Thus, when using triangular FNs, the weight of criterion j is written as $\tilde{w}_j = (w_j^1, w_j^2, w_j^3)$. Let $\tilde{v}_{ij}$ be the element of the weighted normalized fuzzy decision matrix, $\tilde{V}$, corresponding to the pair alternative i, criterion j:

$$\tilde{v}_{ij} = \tilde{w}_j (\cdot) \tilde{r}_{ij}$$

where $(\cdot)$ is the product of two FNs, which has been defined as

$$\tilde{v}_{ij} = (\tilde{w}_j^1, \tilde{w}_j^2, \tilde{w}_j^3)(\cdot)(\tilde{a}_{ij}, \tilde{b}_{ij}, \tilde{c}_{ij}) = (\tilde{w}_j^1 \tilde{a}_{ij}, \tilde{w}_j^2 \tilde{b}_{ij}, \tilde{w}_j^3 \tilde{c}_{ij}).$$

with $\tilde{r}_{ij} = (\tilde{a}_{ij}, \tilde{b}_{ij}, \tilde{c}_{ij})$.

In the example, suppose the DM evaluates the first criterion as very important, the second criterion as important and the last one with low importance. These linguist variables have been modelled by the triangular FN in Table 6.18.

Table 6.18: Linguist Variables Criterion Weight

Very low	(0, 0, 0.2)
Low	(0.1, 0.3, 0.5)
Important	(0.4, 0.6, 0.9)
Very important	(0.6, 0.8, 1)

The fuzzy weighted normalized elements for alternative A are

$$\tilde{v}_{A1} = (0.6, 0.8, 1)(\cdot)(0.9, 1, 1) = (0.54, 0.8, 1),$$

$$\tilde{v}_{A2} = (0.4, 0.6, 0.9)(\cdot)(0.17, 0.33, 0.5) = (0.068, 0.198, 0.45),$$

$$\tilde{v}_{A3} = (0.6, 0.8, 1)(\cdot)(0.1, 0.3, 0.5) = (0.06, 0.24, 0.5).$$

Table 6.19 shows the fuzzy weighted normalized decision matrix, where the first row corresponds to the values computed above.

Table 6.19: Fuzzy Weighted Normalized Decision Matrix

	C_1	C_2	C_3
A	(0.54, 0.8, 1)	(0.08, 0.24, 0.54)	(0.06, 0.24, 0.5)
B	(0.36, 0.64, 1)	(0.32, 0.54, 0.9)	(0, 0.06, 0.2)
C	(0.18, 0.44, 0.8)	(0.2, 0.42, 0.81)	(0.09, 0.3, 0.5)
D	(0.36, 0.64, 1)	(0.04, 0.12, 0.27)	(0.06, 0.24, 0.5)
E	(0.54, 0.8, 1)	(0.04, 0.12, 0.27)	(0.03, 0.165, 0.4)

Step 3F: Fuzzy Ideal and Nadir Solutions

The approach used when computing the fuzzy weighted normalized decision matrix assures us that all $\tilde{v}_{ij}$, $i = 1, ..., n$, $j = 1, ..., k$ vary in $[0,1]$. Therefore, Chen (2000) proposes the fuzzy ideal solution $(\tilde{A}^*)$ and the fuzzy nadir solution $(\tilde{A}^-)$ to be defined as

$$\tilde{A}^* = (\tilde{v}_1^*, \tilde{v}_2^*, ..., \tilde{v}_k^*) \text{ and } \tilde{A}^- = (\tilde{v}_1^-, \tilde{v}_2^-, ..., \tilde{v}_k^-)$$

where $\tilde{v}_j^* = (1, 1, 1)$ and $\tilde{v}_j^- = (0, 0, 0)$, $j = 1, ..., k$.

Note once again, since in the fuzzification step, the cost criterion was transformed into a benefit criterion, then the best value it attains is 1. If this step has not been taken (e.g. the same linguist variable had been used for both benefit and cost criteria), then the normalization step would have transformed the cost criterion into a benefit one. And so, the best value would still be 1.

Step 4F: Compute the Distance from Fuzzy Ideal and Nadir Solutions

The method proposed to calculate the distance between two triangular FNs is the vertex method. Given two triangular FNs, $\tilde{a} = (a_1, a_2, a_3)$ and $\tilde{b} = (b_1, b_2, b_3)$, the distance between them is given by

$$d(\tilde{a}, \tilde{b}) = \sqrt{\frac{1}{3}[(a_1 - b_1)^2 + (a_2 - b_2)^2 + (a_3 - b_3)^2]}$$

and the distances between each alternative i, $i = 1, ... n$, and $\tilde{A}^*$ and $\tilde{A}^-$ are defined, respectively, as

$$d_i^* = \sum_{j=1}^{k} d(\tilde{v}_{ij}, \tilde{v}_j^*),$$

$$d_i^- = \sum_{j=1}^{k} d(\tilde{v}_{ij}, \tilde{v}_j^-).$$

Let's determine the distance of alternative A from $\tilde{A}^*$. Since following the FTOPSIS method first proposed by Chen, the fuzzy ideal solution will have its triplet as $\tilde{v}_j^* = (1, 1, 1)$, $j = 1, ..., k$. Then,

$$d(\tilde{v}_{A1}, \tilde{v}_1^*) = \sqrt{\frac{1}{3}[(0.54 - 1)^2 + (0.8 - 1)^2 + (1 - 1)^2)]} = 0.2896$$

$$d(\tilde{v}_{A2}, \tilde{v}_2^*) = \sqrt{\frac{1}{3}[(0.08 - 1)^2 + (0.24 - 1)^2 + (0.54 - 1)^2)]} = 0.738$$

$$d(\tilde{v}_{A3}, \tilde{v}_3^*) = \sqrt{\frac{1}{3}[(0.06 - 1)^2 + (0.24 - 1)^2 + (0.5 - 1)^2)]} = 0.7552 .$$

Thus, $d_A^* = 0.2896 + 0.738 + 0.7552 = 1.7828$. Similar operations allow for the calculation of the distance from the fuzzy nadir solution (d_i^-). Table 6.20

Table 6.20: Distance from Fuzzy Ideal Solution (d^*) and from the Nadir Solution (d^-)

	d^*	d^-
A	1.783	1.468
B	1.818	1.471
C	1.889	1.417
D	2.041	1.21
E	1.968	1.224

shows the distance of all alternatives from the fuzzy ideal solution (d^*) and from the fuzzy nadir solution (d^-).

Steps 5F and 6F: Compute the Closeness Coefficient and Rank the Alternatives

The closeness coefficient (CC^*) for the FTOPSIS is the same as the relative closeness coefficient proposed for TOPSIS. This coefficient aggregates the two previous distances as

$$C_i^* = \frac{d_i^-}{d_i^* + d_i^-}, i = 1, ..., n.$$

Again, its definition assures $0 < CC_i^* < 1, i = 1, ..., n,$ and hold the property of $CC_i^* = 1$ when $i = \tilde{A}^*$ and $CC_i^* = 0$ when $i = \tilde{A}^-$. So, the highest the value of CC_i^*, the closer the alternative is to the ideal solution. The ranking of the alternatives can be established by sorting $CC_i^*, i = 1, ..., n$ in decreasing order.

Table 6.21 shows the closeness coefficient for all five alternatives and the final ranking:

$$1^{st} - A; 2^{nd} - B; 3^{rd} - C; 4^{th} - E; 5^{th} - D.$$

Table 6.21: Closeness Coefficient (CC^*)

	CC^*	Rank
A	0.4516	1
B	0.4472	2
C	0.4286	3
D	0.3722	5
E	0.3835	4

6.4.3   Fuzzy TOPSIS for Group Decision Making

Suppose that there is a group of decision makers associated with one MCD problem. Each DM will evaluate the alternatives according to each criterion and will define his/her relative importance weights. Chen (2000) was the first to address this topic using FTOPSIS. Since then, alternative ways to deal with group decision making and FTOPSIS have been developed. The method described below, follows Chen's (2000) work. The DM's different evaluations and weights are aggregated prior to the application of the 6step approach.

Before presenting the aggregation method, one needs to define two arithmetic operations involving triangular FNs. Given two triangular FNs $\tilde{a} = (a_1, a_2, a_3)$ and $\tilde{b} = (b_1, b_2, b_3)$ and $\beta \in R$:

- $\tilde{a} (+) \tilde{b} = (a_1 + b_1, a_2 + b_2, a_3 + b_3)$
- $\beta\tilde{a} = (\beta a_1, \beta a_2, \beta a_3)$.

The aggregation step is quite straightforward, once all linguist variables used by each DM have been translated into fuzzy numbers. Let D be the number of Decision Makers, and let $\tilde{x}_{ij}^D$ and $\tilde{w}_j^D$ be the rating FNs and the weight FN of the Dth decision maker, respectively. The aggregated rating and weight FNs are given by

$$\tilde{x}_{ij} = \frac{1}{D}[\tilde{x}_{ij}^1(+) \, \tilde{x}_{ij}^2(+) \dots (+) \, \tilde{x}_{ij}^D]$$

$$\tilde{w}_j = \frac{1}{D}[\tilde{w}_j^1(+) \, \tilde{w}_j^2(+) \dots (+) \, \tilde{w}_{ij}^D]$$

After computing these FNs, the FTOPSIS proceeds as described in the previous section.

Suppose in the example, the team leader and two other team collaborators will evaluate the candidates (DM_1, DM_2, DM_3). When asked to evaluate the importance of three criteria (C_1, C_2 and C_3) using the linguist variable defined in Table 6.18, each of them have classified them differently (Table 6.22).

Table 6.22: Criteria Importance Evaluation by the Three DMs

	DM_1		DM_2		DM_3	
C_1	Very important	(0.6,0.8,1)	Important	(0.4,0.6,0.9)	Very Important	(0.6,0.8,1)
C_2	Important	(0.4,0.6,0.9)	Important	(0.4,0.6,0.9)	Very Important	(0.6,0.8,1)
C_3	Low	(0.1,0.3,0.5)	Very Low	(0,0,0.2)	Low	(0.1,0.3,0.5)

The triangular FN representing the aggregated weighs to be used in step 2F is $\tilde{W} = (\tilde{w}_1, \tilde{w}_2, \tilde{w}_3)$ with

$$\tilde{w}_1 = \frac{1}{3}[(0.6, 0.8, 1)(+)(0.4, 0.6, 0.9)(+)(0.6, 0.8, 1)] = (0.533, 0.733, 0.967),$$

$$\tilde{w}_2 = \frac{1}{3}[(0.4, 0.6, 1)(+)(0.4, 0.6, 0.9)(+)(0.6, 0.8, 1)] = (0.467, 0.667, 0.933),$$

$$\tilde{w}_3 = \frac{1}{3}[(0.1, 0.3, 0.5)(+)(0, 0, 0.2)(+)(0.1, 0.3, 0.5)] = (0.033, 0.2, 0.4).$$

Similarly, the three fuzzy decision matrices should be aggregated into a single fuzzy decision matrix. With these two (prior) steps taken, one can proceed with the FTOPSIS methods starting at step 1F.

As always, this simple aggregation approach is not without some criticisms. Mahdavi et al. (2008) proposed a different fuzzy TOPSIS model and compared the results of the numerical example when applying three different fuzzy TOPSIS methods, amongst them the one proposed by Chen (2000). The authors concluded that "the results indicate that the correct solution is found only by our method, and the other three methods obtain an incorrect result". Huang and Li (2012) point out that Chen's method does not take into account the preferences of individual DMs with respect to the alternatives. As a consequence, the final ranking may lack consensus among the DMs and problems may appear when implementing the chosen decision. The authors propose an alternative approach and argue that it leads to "a decision which is more realistic and acceptable for decision makers".

Although presented in a fuzzy environment of decision making, there are several works that address group MCD making in non-fuzzy context. The interested reader is referred to Shih et al. (2007) where a (non-fuzzy) group decision TOPSIS method is proposed.

6.5 Final Remarks

Just two final remarks to finish this chapter. The first is that shortly after being published, TOPSIS was extended to address multi-objective decision making problems (Lai et al., 1994). These are problems with an infinite number of alternatives (defined by a set of inequalities) that are evaluated by a finite number of criteria (known as "objective functions"). Multi-objective decision making is addressed in the next chapters of this book. The second is to mention that Python implementations for both the initial version of TOPSIS and for Fuzzy TOPSIS for group decision making are available in Papathanasiou and Ploskas (2018).

6.6 Proposed Exercises

1. Gabriela wants to buy a new car and does not know where to begin. She is a very good friend of yours and you are willing to help her. The first question you asked her was: "What points of view do you consider important in a car?" After some thinking, Gabriela told you that:

"For me, the most important aspect is how it looks! I want people to watch and say: 'What a nice set of wheels she is driving'. Secondly, the price, the lower, the merrier. More money for the fuel. Finally, the car's performance. The price is a bit less important than the looks, but the performance, although I found it relevant, is much less important than the other two issues."

Using a magazine, you collected the information and organized it in Table 6.23.

Unfortunately, Gabriela has a budget of 1,000 Plins, consequently some cars had to be removed from Gabriela's initial selection. The cars' performances were taken from the magazine's evaluations on a scale from 0 (the worst) to 20 (the

Table 6.23: Information concerning Six Car Alternatives Evaluated in Criteria Price, Looks and Performance

Car	Price (in Plins)	Looks	Performance
A	350	80	12
B	750	80	18
C	750	100	20
D	120	0	15
E	250	60	12
F	350	60	11

best). The "looks" were classified by Gabriela in a scale from 0 to 100. Which car would you recommend to your friend? Perform a sensitivity analysis on the weight of the most important criterion.

2. Brutus, the brave and bold warrior earned a couple of silver pieces on his last adventure. He successfully saved the (not so) beautiful Princess Blimunda from the dreadful Dark Lagoon Monster. Unfortunately, the creature's acid destroyed Brutus' keen blade, Valeria. Back to the city, Brutus went to his friend Ferreira, the blacksmith, to buy a new weapon. Ferreira has some magic weapons available which have been evaluated by Brutus in three characteristics: Price, Damage and Ease of use. All information has been gathered in Table 6.24.

Table 6.24: Evaluation of Ferreira's Available Weapons in Criteria Price, Damage and Ease of Use

Weapon	Price (in silver pieces)	Damage (in hit points)	Easiness of Use
Alice (dagger)	2	4	20
Butch (mace)	5	6	18
Crunch (morningstar)	8	8	13
Dalila (spear)	6	6	13
Erving (sword)	15	8	10
Fandango (axe)	20	12	5

Brutus does not intend to buy a weapon that does less than 5 hit points. He has a very linear mind and considers all the mentioned attributes equally important. Help Brutus pick his new weapon of choice. Perform a Sensitive Analysis to evaluate your suggestion's robustness.

3. Evarista, the Witch, has a problem. She has to run away from her coven's town because the other witches discovered she is doing good deeds for the town folks. When she applied for the place, she had no idea it was an Evil Coven! And now the other crones are out there to get her. Fortunately, she has been doing this witch's life for decades and she is prepared for these kinds of circumstances. She has several caches in different towns so she can start again somewhere else. The problem is, escaping is risky, and she does not know what town she prefers to escape to. After she gathered her thoughts, she settled the following 6 towns that were assessed according to cache saved, probability of getting caught and town's pleasantness (Table 6.25).

Table 6.25: Information collected by Evarista Concerning Possible Escape Towns.

Town	Cache (in Gold Pieces)	Probability of Getting Caught	Town's Pleasantness
Altamira	700	0.05	18
Bayona	1500	0.25	5
Cantlecaster	1300	0.25	0
Doh	240	0.00	10
Eboricum	1600	0.55	18
Ferim	500	0.10	18

Evarista does not want to take too high a risk. She will not try to escape to a town with a probability of being caught greater than 30%. The Town's pleasantness was measured in scale from 0 (very bad) to 20 (excellent) points. Recommend a destiny to Evarista considering equal weights for the three criteria. Perform a sensitivity to a criterion of your choice to assess your suggestion's robustness.

4. Auntie Leonarda needs to hire a private jet plane for crossing the continent and go to an important event, tomorrow. She checked a site on-line and pre-selected the four models that pleased her most. Table 6.26 presents all information Auntie Leonarda considers relevant for her final decision.

Table 6.26: Information Gathered by Auntie Leonarda

Jet Model	Cost (in Monetary Units) C_1	Trip Duration (in Hours) C_2	Comfort C_3
Amethyst (A)	3000	6 h	80
Beryl (B)	1500	6 h 30 min	50
Crystal (C)	1000	10 h	60
Diamond (D)	5000	4 h 15 min	100

The Attribute "Comfort" was assessed in points from 0 to 100 (the higher the points the better). Rank the four models knowing that Auntie Leonarda considers "Comfort" the most important criterion, immediately followed by "Trip Duration". "Cost" is considerably less important than the other two. Perform a sensitivity analysis to the most important criterion.

5. Auntie Leonarda is planning her summer vacations and although she has not decided where to go, she knows she will be taking a nephew or niece with her. She wants to take Alzira, but she must "scientifically" justify her decision, or she will have a jealousy riot among the nephew gang. Therefore, she decided to consider the following aspects:

- The nephew/niece's last semester grades (the higher, the better);
- The number of trips each youth has already done with her (the lower, the better. Balancing karma, my dears);
- The value of the gift each of them gave Auntie on her last Birthday (the higher the better... Time for paying back.)

Table 6.27 presents the information collected by Auntie.

Table 6.27: Information Collected by Auntie Leonarda

Nephew/Niece	Grades Average (0–20)	# of Previous Trips	Gift Value (in Plins)
Alzira	15.6	3	10,500
Bernardo	16.2	5	9,000
Cecília	14.3	2	6,500
Duarte	16.0	5	7,500
Ermelinda	15.1	3	10,000

Obtain the final ranking assuming equal weights. Considering one criteria of your choice, investigate whether Alzira can ever be ranked 1st.

6. Jill and Jane (J & J) are buying a house. After a long market consultation, they selected the most interesting options. All have similar areas and building quality. The choice will be made based on:

- Price, in plins;
- Average distance from their workplaces, in km;
- Pleasantness, evaluated using a 0–20 scale.

J&J organized all relevant information in Table 6.28

Table 6.28: Information Collected by J&J

Alternative	Price	Distance	Pleasantness
W	3,000	30	16
X	5,000	10	14
Y	9,000	15	20
Z	9,000	50	13

J&J consider the distance to their workplaces the most important criterion. Price and Pleasantness are equally important, but less than the Distance criterion. Obtain the final ranking. Analyze the ranking's robustness.

7. The Government of Lusoland is considering the viability of building a third bridge across river Tajus in Lisbonia metropolitan area. There are four possible places where the bridge can be built. The four alternatives were evaluated according to three points of view: Cost (in millions of Eurons); Environmental Impact (in a scale of 0–null to 20–catastrophic); and Road Traffic relief (in percentage).

The information concerning the evaluations was collected and is presented in Table 6.29.

Table 6.29: Information gathered Concerning Four Possible Crossing Locations

Possible crossing	Cost	Environmental Impact	Traffic Relief
Arreiro – Lisbonia	300	9	55%
Brandão – Lisbonia	150	6	50%
Ceixal – Lisbonia	100	5	5%
Drafaria – Lisbonia	500	18	90%

For political reasons, the Government intends to build the bridge connecting Drafaria to Lisbonia. Knowing the Government cannot ignore the cost and environmental impact criteria, perform a sensitivity analysis to the weight of "traffic relief" to investigate whether this alternative will ever be ranked 1st.

8. Auntie Leonarda wants to throw a tremendous party to celebrate the beginning of the Social Season. Zumélia, her niece and assistant, suggested some possible themes for the party: the Aqua Party (a traditional celebration by the artificial lake in Auntie's mansion grounds, lots of fountains and waterfalls); the Black Party (everyone dressed in elegant black clothes, dark decoration and black food and beverages); the China Party (Far East themed decoration, Oriental inspired music and clothes, Chinese food); the Divine Party (guests must dress like Greek or Egyptian Mythology characters, ethereal and grandiose decoration). To help Auntie Leornarda, Zumélia asked the right questions and was able to fill out Table 6.30, with three criteria's perspective: cost, positive impact on the jet set society and organization difficulty.

Table 6.30: Information Organized by Zumélia

Party	Cost	Jet Set	Difficulty
Aqua	Very high	High	Very high
Black	Average	Low	Null
Chinese	Very low	Average	Very Low
Divine	Extreme	Extreme	Average

As Zumélia knows Auntie Leornarda since forever, she is sure that the triangular FNs in Table 6.31 adequality model Auntie's evaluation of each of the adjectives she used.

Table 6.31: Triangular FN

Null	(0, 0, 1)
Very low	(1, 2, 3)
Low	(2, 4, 6)
Average	(4, 7, 10)
High	(13, 11, 13)
Very high	(13, 15, 18)
Extreme	(17, 18, 18)

To Auntie Leornarda, the positive impact in the jet set socialite is, undoubtedbly, very important. The cost is important but the difficulty to put up a party is very low. Help Zumélia to rank the four alternatives.

9. The Mayor of Vila Nova de Carambola needs to build a new landfill in the municipality. Obviously, whatever place is chosen will cause protest and loss of popular support. So, the decision has to be carefully taken. There are four available sites that were already evaluated from three points of view using linguist variables: distance to the main town (the further, the better... out of sight, out of mind...), positive impact on the next elections, and environmental impact. The evaluations' results can be seen on Table 6.32.

Table 6.32: Evaluation of Four Possible Sites to Build the Vila Nova de Carambola New Landfill

Site	Distance	Elections	Env. Impact
W	Far	Low	High
X	Close	Very high	Low
Y	Too close	Average	Low
Z	Very far	Very low	Very high

After a deep reflection, the Mayor was able to define two triangular FNs related with the linguist variables that he has used. Table 6.31 shows the triangular FNs for the linguist variables used on criteria "elections" and "environmental impact", while Table 6.33 presents the one for the "distance" criterion.

Table 6.33: Triangular FN for Criterion "Distance"

Too close	(0, 0, 4)
Close	(0, 4, 8)
Halfway	(6, 12, 16)
Far	(12, 16, 20)
Very far	(17, 20, 20)

The Mayor considers the impact on the elections a very important criterion and the other two equally important. Help the Mayor pick the new location.

References

Bellman, R.E. and L.A. Zadeh (1970). Decision-making in a fuzzy environment. *Management Science*, 17(4), B–141.

Chen, C.T. (2000). Extensions of the TOPSIS for group decision-making under fuzzy environment. *Fuzzy Sets and Systems*, 114(1), 1–9.

Chen, S.J. and C.L. Hwang (1992). Fuzzy multiple attribute decision making methods. *Fuzzy Multiple Attribute Decision Making*, 289–486.

Cheng, C.B. (2004). Group opinion aggregation based on a grading process: A method for constructing triangular fuzzy numbers. *Computers & Mathematics with Applications*, 48(10-11), 1619–1632.

de Farias Aires, R.F. and L. Ferreira (2019). A new approach to avoid rank reversal cases in the TOPSIS method. *Computers & Industrial Engineering*, 132, 84–97.

García-Cascales, M.S. and M.T. Lamata (2012). On rank reversal and TOPSIS method. *Mathematical and Computer Modelling*, 56(5-6), 123–132.

Hanss, M. (2005). *Applied Fuzzy Arithmetic*. Springer-Verlag Berlin Heidelberg.

Huang, Y.S. and W.H. Li (2012). A study on aggregation of TOPSIS ideal solutions for group decision-making. *Group Decision and Negotiation*, 21(4), 461–473.

Hwang, C.L. and K. Yoon (1981). Methods for multiple attribute decision making. *In*: Multiple Attribute Decision Making (pp. 58–191). Springer, Berlin, Heidelberg.

Ishizaka, A. and P. Nemery (2013). *Multi-criteria Decision Analysis: Methods and Software*. John Wiley & Sons.

Klir, G. and B. Yuan (1995). *Fuzzy Sets and Fuzzy Logic* (Vol. 4). New Jersey: Prentice hall.

Lai, Y.J., T.Y. Liu and C.L. Hwang (1994). TOPSIS for MODM. *European Journal of Operational Research*, 76(3), 486–500.

Nădăban, S., S. Dzitac and I. Dzitac (2016). Fuzzy TOPSIS: A general view. *Procedia Computer Science*, 91, 823–831.

Olson, D.L. (2004). Comparison of weights in TOPSIS models. *Mathematical and Computer Modelling*, 40(7-8), 721–727.

Opricovic, S. and G.H. Tzeng (2004). Compromise solution by MCDM methods: A comparative analysis of VIKOR and TOPSIS. *European Journal of Operational Research*, 156(2), 445–455.

Palczewski, K. and Salabun (2019). The fuzzy TOPSIS applications in the last decade. *Procedia Computer Science*, 159, 2294–2303.

Papathanasiou, J. and N. Ploskas (2018). Topsis. *In*: Multiple Criteria Decision Aid (pp. 1–30). Springer, Cham.

Ribeiro, R.A. (1996). Fuzzy multiple attribute decision making: A review and new preference elicitation techniques. *Fuzzy Sets and Systems*, 78(2), 155–181.

Ribeiro, R.A., A. Falcao, A. Mora and J.M. Fonseca (2014). FIF: A fuzzy information fusion algorithm based on multi-criteria decision making. *Knowledge-Based Systems*, 58, 23–32.

Salih, M.M., B.B. Zaidan, A.A. Zaidan and M.A. Ahmed (2019). Survey on fuzzy TOPSIS state-of-the-art between 2007 and 2017. *Computers & Operations Research*, 104, 207–227.

Shih, H.S., H.J. Shyur and E.S. Lee (2007). An extension of TOPSIS for group decision making. *Mathematical and Computer Modelling*, 45(7-8), 801–813.

Vafaei, N., R.A. Ribeiro and L.M. Camarinha-Matos (2018). Data normalisation techniques in decision making: Case study with TOPSIS method. *International Journal of Information and Decision Sciences*, 10(1), 19–38.

Wang, X., D. Ruan and E.E. Kerre (2009). *Mathematics of Fuzziness—Basic Issues* (Vol. 245). Springer.

Wang, Y.M. and Y. Luo (2009). On rank reversal in decision analysis. *Mathematical and Computer Modelling*, 49(5-6), 1221–1229.

Yatsalo, B., A. Korobov and L. Martinez (2021). From MCDA to Fuzzy MCDA: Violation of basic axiom and how to fix it. *Neural Computing and Applications*, 33(5), 1711–1732.

Zadeh, L.A. (1975). The concept of a linguistic variable and its application to approximate reasoning – I. *Information Sciences*, 8(3), 199–249.

Multi-Objective Linear Programming

7.1   Introduction

When making choices one is faced with different ways of looking at the object of decision and one wishes to make the best choice with respect to each one of them. However, this is not always possible. In fact, it is rare that those perspectives lead to the same "best" option. In the previous chapters in this book, several methods have been presented addressing problems where the Decision Marker (DM) has a finite number of alternatives and aims at selecting the best one considering several points of view. In this chapter, one will focus on methodologies that can be used when alternatives are infinite in number. In fact, alternatives are not explicitly defined but rather implicitly defined by a set of constraints. Since the decisions need to account for multiple perspectives, this chapter addresses what is known as Multi-Objective Linear Programming (MOLP) problems.

MOLP is, within Mathematical Programming, the area that tackles linear optimization models with multiple objectives. The importance of this optimization approach is shown by the large spectrum of applications it has been applied to. Sustainability, for instance, is multi-objective in nature since sustainable decisions should simultaneously account for the maximization of economic throughput, the minimization of environmental impact and the maximization of social benefit. Many other applications have been showing up in literature. Anderluh et al. (2020) studied the delivery problem where inner-cities vehicles (smaller vehicles) need to be routed in synchronization with conventional delivery vehicles so that total transportation costs, GHG emission and inner-cities traffic disturbance are minimized. Javanmard et al. (2020) make use of multi-objective optimization to optimize a buildings' energy and water consumption and reduce CO_2 emissions. Javid et al. (2020) addressed batch production systems accounting for both the trade-off between flexibility and complexity through the modeling of seven objectives to be optimized. These are just three very recent examples of multi-objective optimization modelling in sustainability planning. Other applications can be found in Medicine (e.g, Van Haveren et al. (2020) developed an automated treatment planning algorithm for oropharyngeal cancer patients considering 22 objectives); in portfolio selection (e.g. Ruiz et al. (2020) developed a new credibility portfolio selection model taking into consideration three perspectives: credibilistic expected value, a risk measure and a loss function); in classification problems (e.g. Ribeiro and Reynoso-Meza (2020) study multi-objective approaches to build ensemble models for problems of learning on imbalanced data sets) just to name a few.

The body of literature that makes use of multi-objective optimization is massive which shows how fundamental this topic is when solving a real-world problem with all its complexities. The terminology used in the MOLP can be confusing since there are different ways of defining the MOLP main concepts (see Ehrgott 2005). This chapter will focus on the fundamental topics and concepts in MOLP, which will be used throughout the remaining chapter of this book. It will focus exclusively on linear models. For mixed-integer and non-linear multi-objective problems, the interested reader should refer to reference books such as Miettinen (1998) or Antunes et al. (2016).

Generically, a MOLP model can be written on the form:

$$\text{Max } Z_1 = \sum_{i=1}^{n} c_{i1}x_i$$

$$\text{Max } Z_2 = \sum_{i=1}^{n} c_{i2}x_i$$

...

$$\text{Max } Z_k = \sum_{i=1}^{n} c_{ik}x_i$$

$$\text{s.t. } \sum_{i=1}^{n} A_{ij}x_i \geq b_j, \quad \forall j = 1,...,m$$

$$x_i \geq 0, \quad \forall i = 1, ..., n$$

where c_{ip}, $i = 1,..., n$, $p = 1,..., k$ are the coefficients of the objective functions, A_{ij} is the $n . m$ matrix with constraints coefficients and b_j, $j = 1,..., m$ are the constant terms on the right side of the constraints. This problem is comprised of n variables, m constraints and k objective functions.

In more compact form one can define a MOLP model as

$$\text{Max } (Z_1(x), Z_2(x), ..., Z_k(x))$$
$$\text{s.t. } x \in S$$

with x the vector of decision variables defined over S, the feasible region defined by all model constraints, and $Z_1(x), Z_2(x), ..., Z_k(x)$ the k objective functions.

Throughout this chapter, the feasible region S is assumed to be a non-empty set and all objective functions attain a maximum value in the feasible region. Without loss of generality, the maximization is considered for each objective function. Nonetheless, problems exist where some or all objective functions are to be minimized. In such cases, the traditional transformations can be applied to rewrite a minimization objective problem as a maximization one. In section 7.6, an exercise with one objective function as a minimization objective is solved. Lastly, in the literature, one often find models with two objective functions ($k = 2$) named as a biobjective problem. In this book no such distinction will be made.

7.2   Efficient Solutions and Objective Functions Space

The conflicting nature of the objective functions in a MOLP problem makes it impossible to find one solution that optimizes all objective functions. This renders meaningless the concept of optimal solution which is the one of the most important concepts in single objective optimization[1]. Therefore, when multiple objectives are modelled one aims at determining a solution that yields the best compromise amongst all objective functions. These solutions are known as efficient solutions and are the ones that may be of interest to the Decision Maker (DM).

More precisely, given a feasible space S, a **solution** $x^* \in S$ is **efficient** if there exists no other solution $x \in S$ that would improve some objective function without simultaneously worsening at least one other objective function. Formally, one may define efficient solution as:

Definition of Efficient Solution

A solution $x^* \in S$ is called efficient if and only if there is no other solution $x \in S$ such that

(i) $Z_i(x) \geq Z_i(x^*)$, $i = 1, \ldots, k$

(ii) $\exists_{i = 1, \ldots, k} Z_i(x) > Z_i(x^*)$.

Let's take a small example with two objective functions so that one can visualize the geometry behind this concept.

Example 7.1

Consider the problem where the Decision Maker wishes to find the solution that maximizes Z_1 and Z_2 subjected to four inequalities (forming set S).

Figure 7.1 shows set S, the feasible region defined by the four constraints. It also depicts the two objective functions and the corresponding gradient vectors. When solved independently[2], function Z_1 attains its optimal value of $Z_1^* = 1200$ at vertex (40, 40), while function Z_1 at vertex (0, 40) with an optimal value of Z_2^* = 1600. These two objective functions are conflicting objectives since they do not share the optimal solution. Thus, if the Decision Maker picks any of these two solutions, the one chosen will perform poorly on the other objective.

---

[1]   A very interesting piece of work investigating the concept of optimum as a "balance among multiple criteria" can be found in Zeleny (1998).

[2]   For further details on how to graphically solve a linear programming problem, the interested reader should refer to any Operations Research reference book as, for example, Winston (2002).

$$\text{Max } Z_1 = 20x + 10y$$
$$\text{Max } Z_2 = -80x + 40y$$
$$\text{s.t. } x + y \geq 20$$
$$2x - y \leq 40$$
$$-x + y \leq 80$$
$$x + y \leq 80$$
$$x, y \geq 0$$

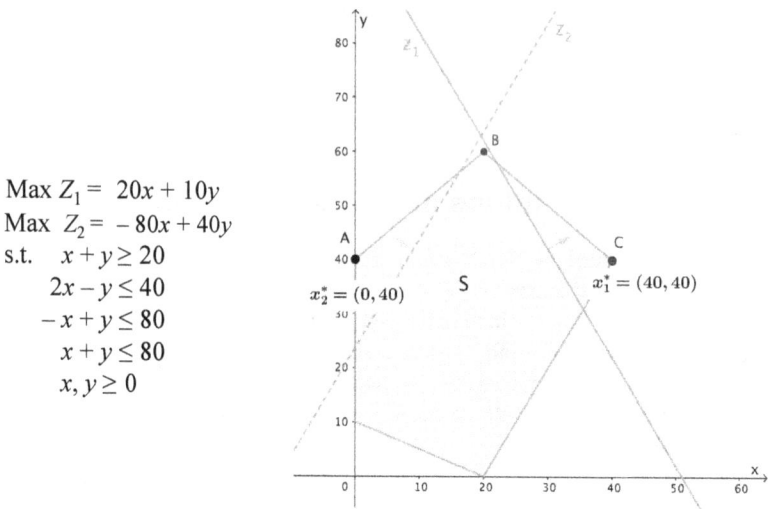

Figure 7.1: Feasible Region S and Optimal Solutions x_1^* and x_2^* for Objectives Z_1 and Z_2, respectively

Solutions $A = (40, 40)$, $B = (20, 60)$ and $C = (40, 40)$ are efficient solutions since no other feasible solution performs better (or equality better) for both objective functions, and strictly better for at least one of the objectives.

But are the extreme points the only efficient solutions? In Fig. 7.2 the region where the solutions that might be of interest to the DM, the set of efficient solutions, is shown. In all three figures, the region where both gradient vectors are pointing inwards is highlighted. This region is known as **dominance cone**[3]. When moving any of the objective functions, the dominance cone will move along with it. If one is able to intersect the dominance cone and the feasible region and get only one point, this point is an efficient solution. Therefore, all solutions in [AB] ∪ [BC] are all efficient solutions (Fig. 7.2). The set of all efficient solutions is known as **Efficient Frontier** and will be represented by S^e.

Figure 7.3 shows two non-efficient solutions: W an interior point of S and U a point in the frontier of S. In both cases, the dominance cone has multiple points in common with the feasible region S. All these common points are solutions that perform better than W or U in at least one (or even in both) objective functions. A visual analysis supported by the dominance cone is only possible in two-dimensional spaces and, with some ingenuity, in three-dimensional spaces. In higher dimensions, since there are no easy visualization methods, this concept is harder to be used to determine all efficient solutions.

In the presence of multiple objective functions, a new space is defined by the images of each feasible solution by each objective function. This space is

---

[3] A formal definition of dominance cone can be found in Steuer (1986).

(a)

(b)

(c)

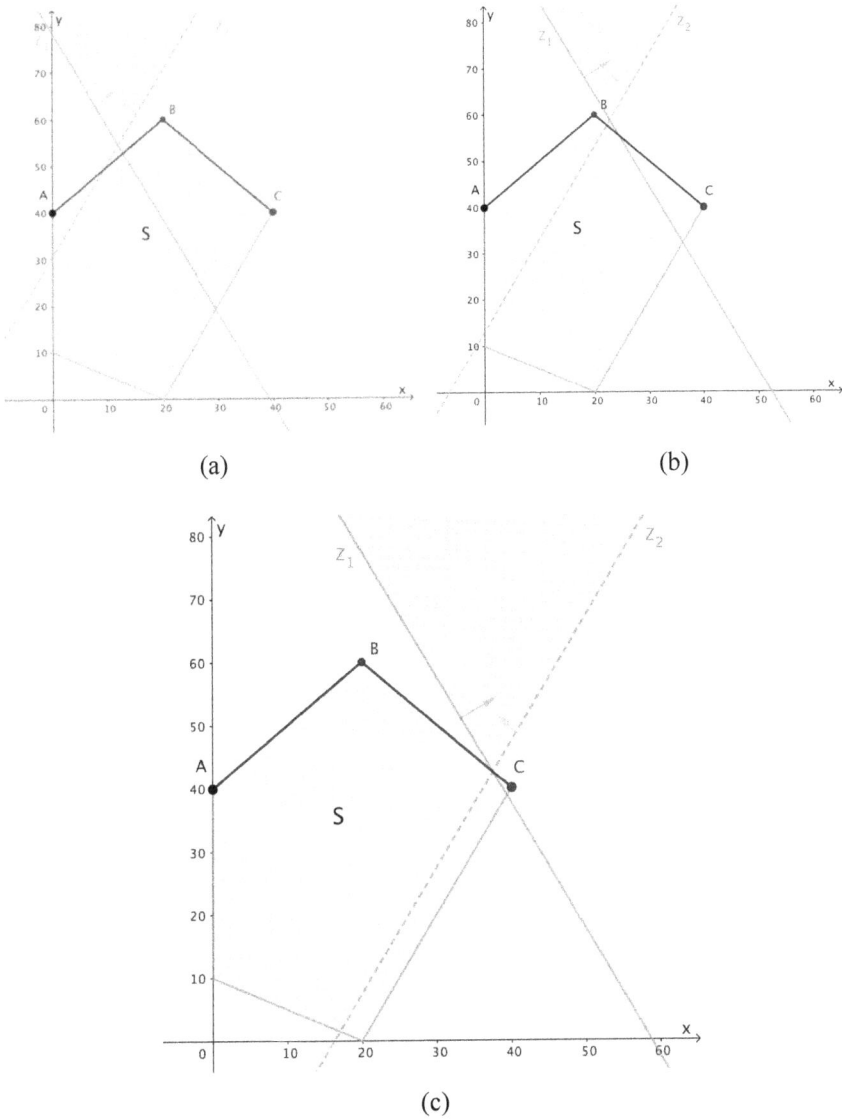

Figure 7.2: Identifying the Efficient Solution in a Two-Dimensional Space with Dominance Cone

known as **objective functions space** or, more simply, as **objectives space** and is represented as S_Z. To each point $x \in R^n$, there is a point $z \in R^k$ such that $z = (Z_1(x), Z_2(x), \ldots, Z_k(x))$. Consequently, to the feasible region $S \subset R^n$ there is a corresponding feasible region in R^k which is the projection of S in R^k by $Z_1(x)$, $Z_2(x), \ldots, Z_k(x)$. Set R^n can also be designated as **solution space** or **decision space**.

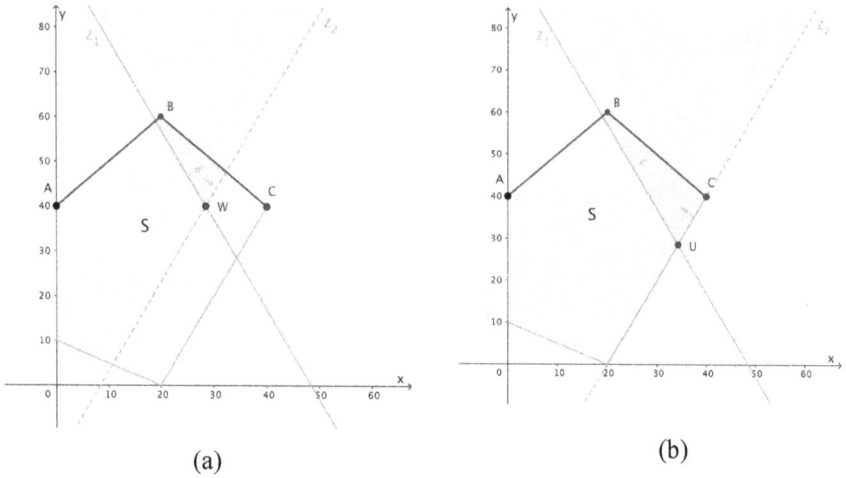

Figure 7.3: Non-Efficient Solution and the Corresponding Dominance Cone in a Two-Dimensional Space

Example 7.2

The feasible region S of Example 7.1 is a polyhedron defined by five vertices. Calculating the images of all vertices of S, one is able to define the corresponding feasible region on the objective function's space.

Vertices in S		Vertices in S_Z	
A	(0,40)	A'	(400,1600)
B	(20,60)	B'	(1000,800)
C	(40,40)	C'	(1200,-1600)
D	(20,0)	D'	(400,-1600)
E	(0,10)	E'	(100,400)

The properties of the objective functions ensure that all vertices in S are mapped into vertices in the objective's space. Hence, the feasible region on the objective functions space can easily be graphically represented (Fig. 7.4). Notice that, although all variables are non-negative, such constraint does not apply to the objective functions' values.

All considerations made above concerning efficient solutions apply to the objective space. Therefore, all solutions in [A' B'] ∪ [B'C'] are the projections of the efficient solutions onto this new space.

Efficient solutions are designated as **non-dominated solutions** when referring to those on the objectives space. Formally, a non-dominated solution may be defined as

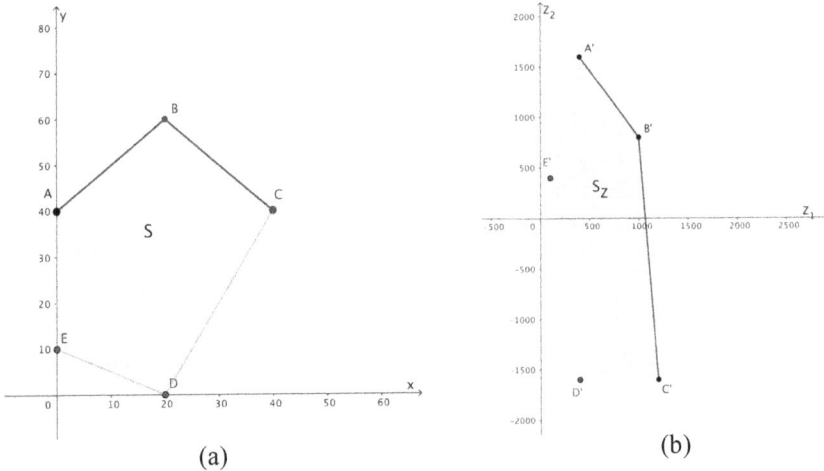

Figure 7.4: Feasible Region Represented on the Solutions Space (a) and on the Objectives Space (b)

Definition of Non-Dominated Solution

A point in the objective functions space $z \in S_Z$ is called a **non-dominated solution** (or **Pareto optimal solution**) if and only if $x \in S$ is an efficient solution with $z = (Z_1(x), Z_2(x), \ldots, Z_k(x))$. In other words,

$$S_Z^e = \{z = (Z_1(x), Z_2(x), \ldots, Z_k(x)): x \in S^e\}.$$

Set S_Z^e is mostly known as **Pareto Front or Pareto Frontier**.

As defined above, the term efficient solution refers to a point in the solutions space and non-dominated solution refers to a point in the objectives space. However, these terms are very often used interchangeably. Similarly, the designation "Pareto Front" may also designate the set of efficient solutions (set S^e) or the set of non-dominated solutions (set S_Z^e).

How to Determine the Pareto Front?

In two dimensional problems (with two variables), it is fairly easy to identify the Pareto Front. However, the two variables problems are only of pedagogical interest. No one has a "problem" with 2 variables. With a larger number of variables, if the problem has two objectives, again it may be quite easy to determine the Pareto Front. Most often, however, problems have a very large number of variables and some may have three or more objective functions. For those problems, several methods have been proposed to compute all efficient vertex of the feasible region (also known as extreme points). All those methods enter into theoretical details concerning Linear Optimization (e.g. efficient basic solutions, efficient extreme rays, among others) that will not be addressed in this book.

The most common approach is to approximate the Pareto Front computing some non-dominated solutions. In the next chapters of this book, a few methods will be studied that allow such approximation.

7.3   Weakly Efficient Solution

An important concept that very often is disregarded by those starting on multi-objective linear optimization is the concept of the weakly efficient solution. This is a more relaxed definition of efficient solutions. A feasible solution is said to be weakly efficient if and only if there is no other feasible solution that strictly improves all objective function values. In order words, a feasible solution is weakly efficient if the condition

$$\exists_{i=1, ..., k} Z_i(x) > Z_i(x^*)$$

is not verified.

Definition of Weakly Efficient and Weakly Non-Dominated Solutions

A solution $x^* \in S$ is said to be **weakly efficient** if and only if there is no other solution $x \in S$ such that

$$Z_i(x) > Z_i(x^*), \ i = 1, ..., k.$$

Similarly, there is the concept of the weakly non-dominated solution which comes from transporting the definition to the objective function space. Hence, $z = (Z_1(x), Z_2(x), ..., Z_k(x))$ is **weakly non-dominated** if and only if x is weakly efficient.

The definition of weakly efficient solution "contains" the definition of efficient solution. This is to mean that all efficient solutions are also weakly efficient, but not the other way around. There are some efficient solutions that are *strictly* efficient. Though, in this textbook when weakly efficient solutions are mentioned, one is not considering the strictly efficient solutions. Likewise, when referring to efficient solutions, one is excluding the weakly efficient ones.

What is the relevance of these solutions? Weakly efficient solutions are not "as good as" (strictly) efficient solutions. For any weakly efficient solution there is at least one other solution that performs better in one objective function while not worsening the value of the remaining objective functions.

Example 7.3

Consider the MOLP problem below where the feasible region and the optimal solutions for each objective function are depicted in Fig. 7.5.

Max $Z_1 = x + y$

Max $Z_2 = -x + y$
s.t. $x + y \le 25$
      $-x + 2y \le 20$
           $x \le 18$
        $x, y \ge 0$

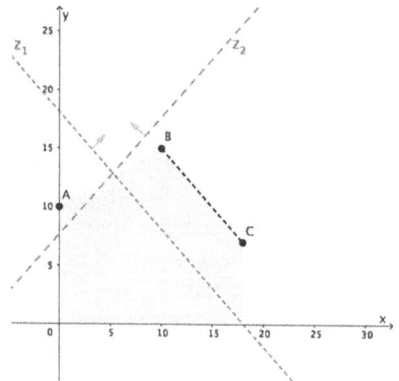

Figure 7.5: Feasible Region and Optimal Solutions for Objectives Z_1 and Z_2.

In this problem, Z_1 is parallel to the constraint $x + y \leq 25$ and so all solutions in segment $[BC]$ are optimal, $Z_1(x) = 25$ for all $x \in [BC]$. Objective function Z_2 attains its optimal value of 10 at vertex $(0, 10)$. Figure 7.6 shows the dominance cone in two different positions. In Fig. 7.6 (a) its vertex is on a point in segment $[AB]$ and it has no other point in common with the feasible region. Moreover, if one positions the dominance cone onto any point in $[AB]$, the intersection between the dominance cone and the feasible region returns only the vertex. Therefore, one can conclude that segment $[AB]$ only contains feasible solutions. Performing a similar analysis on Fig. 7.6 (b), one sees that the intersection of the dominance cone with segment $[BC]$ has more than one point. In fact, part of the segment $[BC]$ is "covered" by one edge of the dominance cone. This means that, with the exception of vertex B, all solution on segment $[BC]$ are weakly efficient. Note that all of them perform equally well when compared to solution B when considering objective function Z_1 but have worse values than B when considering objective function Z_2.

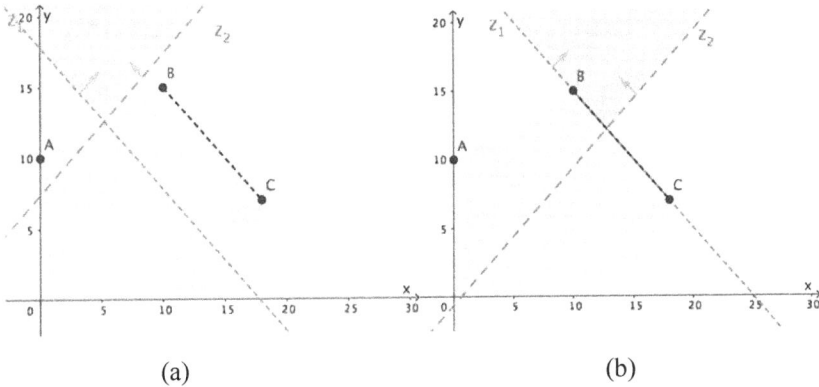

(a)   (b)

Figure 7.6: Dominance Cone Showing (a) Efficient Solutions and
(b) Weakly Efficient Solutions

This is quite evident if one looks into the objectives space (Fig. 7.7). The vertical position of segment $[B'\ C']$ shows that all its solutions attain the same value in Z_1 and different values with respect to Z_2. They are all, with the exception of B', weakly non-dominated solutions. For a rational DM only solutions on segment $[A'\ B']$ are possible compromise solutions.

The major issue with weakly efficiency (or weakly non-domination) is that optimization software is not able to evaluate if an optimal solution is efficient in the strict or in the weak sense. To do so, an algorithm must be implemented to assess the "quality" of the optimal solution with respect to the different objective functions. This topic will be addressed in more detail in Chapter 8 where a method is presented that overcomes this issue.

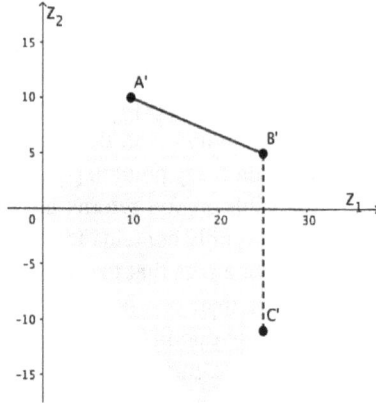

Figure 7.7: Non-Dominated and Weakly Non-Dominated Solutions in the Objective
Functions Space (Example 7.3).

7.4   Payoff Table, and Ideal and Nadir Solutions

The most challenging aspect in MOLP is to determine the best compromise
solution among the Pareto optimal solutions. This challenge will be covered in
the remaining chapters of this book. However, some methods that will be studied
need to access the range of variation of the objective functions over the Pareto
Front. This is to mean that one needs to know the maximum and minimum values
over the efficient set for each objective. While determining the maximum value
over the efficient set is straightforward (one needs only to solve each objective
function individually), to determine the minimum value one needs to solve the
problem

$$\text{Max } Z_p(x)$$
$$\text{s.t. } x \in S^e$$

with $p = 1, ..., k$.

Unfortunately, the solution of this problem is not as easy as the maximum one
since the set S^e is not known explicitly (Steuer, 1986).

The most common strategy is to determine the maximum and minimum
values each objective attains at the optimal solutions of all other objectives. This is
usually done using an instrument named the **payoff table**. This table provides the
objective function values when each objective function is optimized individually:
the first column presents all optimal solutions and the remaining columns the
image of these solutions by each objective function.

Table 7.1 is a generic payoff table where $z_p{}^i$ is the value of the optimal solution
$(x_i{}^*)$ of the objective function Z_i by objective function Z_p. Each row in the payoff
table corresponds to a solution in the objective functions space. The values on the
main diagonal $(Z_p{}^*, p = 1, ..., k)$ are the optimal value of each objective function.

Table 7.1: Generic Payoff Table

	Z_1	Z_2	...	Z_k
x_1^*	Z_1^*	z_1^2	...	z_1^k
x_2^*	z_2^1	Z_2^*		z_2^k
⋮			⋱	
x_k^*	z_k^1	z_k^2		Z_k^*

Two important solutions can be derived from the payoff table. The point in the objectives space $Z^* = (Z_1^*, Z_2^*, ..., Z_k^*)$ is known as the **ideal solution**. This solution is also known as utopia solution or utopia vector. The **nadir solution** is the point whose coordinates are the worst values (the minimum values) for the objective functions over the non-dominated solutions:

$$N^* = (n_1^*, ..., n_k^*)$$

where $n_p^* = \min_{j=1,...,k} z_j^p$, $p = 1, ..., k$. These two solutions are often used to define the range of variation of each objective function over the non-dominated set.

Example 7.4

Table 7.2 is the payoff table for Example 7.1. x_1^* and x_2^* are the optimal solutions for Z_1 and Z_2, respectively. Each row corresponds to the image of each optimal solution in the objective space (see Example 7.2). According to the payoff table, the value range for Z_1 over the non-dominated solutions is [400, 1200], while for Z_2 it is [−1600, 1600].

Table 7.2: Payoff Table for the Problem in Example 7.1

	Z_1	Z_2
$x_1^* = (20, 0)$	1200	−1600
$x_2^* = (0, 40)$	400	1600

The nadir and ideal solutions in this problem are points (-1600,400) and (1200,1600), respectively. In this example the nadir point is a vertex of the feasible region in the objectives space, which is just a coincidence. In fact, the nadir point may fall inside or outside of set S_Z.

Although the payoff table is one of the most use tools to evaluate the range of the objective functions over the Pareto Front (the set of non-dominated solutions), it is not an unproblematic tool. In fact, with the exception of two objective problems, the payoff table too often does not provide the nadir point (Antunes et al. 2016). When problems have multiple optimal solutions, the payoff table may not be unique and consequently, the nadir point may, or may not, be unique.

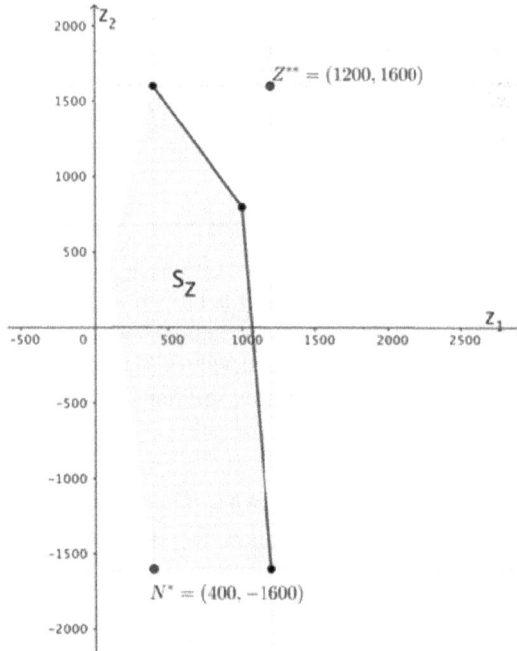

Figure 7.8: Ideal and Nadir Points

Example 7.5

Let's consider the problem in Example 7.3. In this problem, while objective function Z_2 has one single optimal solution, $x_2^* = (0, 10)$, function Z_1 has an infinite number of solutions, all of them corresponding to the optimal value $Z_1^* = 25$. Taking only the extreme points[4], two payoff tables may be computed (Table 7.3).

Table 7.3: Payoff Tables for the Problem of Example 7.3

	Z_1	Z_2		Z_1	Z_2
$x_1^* = (10, 15)$	25	−5	$x_1^* = (18, 7)$	25	−11
$x_2^* = (0, 10)$	10	10	$x_2^* = (0, 10)$	10	10

The ideal solution is (25, 10). The nadir solution may by point (10, −5) or (10, −11). Since this is such a small problem, allowing for a graphical representation, one already knows that solution $x_1^* = (10, 15)$ is the one that should be chosen for the payoff table, as it is the only one that is not weakly non-dominated.

---

[4] Only extreme feasible points are possible optimal solutions calculated by standard optimization software. The study of why only the vertices of feasible regions (also known as extreme points) are of interest, is outside of this book's scope. The interested reader should refer to Steuer (1986) or Winston (2002) for all the details.

However, suppose the problem was of a higher dimension (more variables and more objective functions), the optimal solution for Z_1 determined by a standard optimization software could have been vertex $C = (18,7)$. Consequently, one would conclude that the nadir point would be $(10, -11)$ and the range for Z_1 over the Pareto Front would be $[-11, 10]$, which overestimated the minimum value over the Pareto Front. One already knows this minimum value to be -5.

In the previous example, although there are multiple optimal solutions for one of the objective functions, the payoff matrix is unique since only one of these optimal solutions is in fact a non-dominated one (the remaining are weakly non-dominated). However, other cases exist where there are multiple payoff tables, all with non-dominated optimal solutions.

Example 7.6

Suppose a three objective functions problem where all functions are to be maximised. In addition, suppose Z_1 has only one optimal solution, Z_2 has two vertices as optimal solutions and Z_3 has three vertices that are optimal solutions. All solutions are summarized in Table 7.4.

Table 7.4: Example 7.6 Payoff Table.

	Z_1	Z_2	Z_3
x_1^*	272	50	300
x_2^*	270	295	250
	269	295	255
x_3^*	270	75	300
	268	179	300
	267	167	300

Amongst all the solutions of Table 7.4, only vertex $(267, 167, 300)$ is weakly non-dominated. All other alternative optimal solutions are non-dominated ones. With these solutions, six pay-off tables may be constructed, resulting in six different nadir points. The "true" nadir point is $(268, 50, 250)$. If any other nadir point would have been selected, the range of the different objective functions could have been either overestimated or underestimated.

Several methods have been developed to overcome this issue. Kok and Spronk (1985) propose a procedure to determine the nadir point. The procedure has a tree structure where, in each node, an optimization problem needs to be solved. Therefore, it may be quite computationally intensive just to find the real nadir point. Korhonen et al. (1997) proposed a heuristic based on reference directions starting from a "nadir" solution from a payoff table. Although being a heuristic procedure, the authors argue that the approach is able to find more accurate nadir solutions than using the payoff tables alone and claim that it is easy to implement. More recently, Alves and Costa (2009) proposed an algorithm, based on the weighted sum approach, that is able to compute the exact nadir values. The authors affirm the algorithm can be used in problems with any number of objectives although the computational time increases significantly with the number of objectives (not so much with the number of variables and constraints).

7.5   Final Remarks

This chapter focuses on the basic concepts in Multi-objective Linear Programming. The aim is to make the reader knowledgeable of concepts such as efficient and weakly efficient solutions, non-dominated and weakly non-dominated solutions, objective functions space, Pareto Front and efficient frontier, among others. There are no details concerning methods to reach compromise solutions (the ones the Decision Maker is searching for) nor even methods to approximate the Pareto Front. Such topics will be covered in the remaining chapters of this book.

Traditionally the methods to determine compromise solutions are classified in different categories with respect to the way the information concerning DM preferences is modelled. The "how" and the "when" of this information is extracted from the DM and plays an important role in the methods. Hwang et al. (1980) proposed one of the first taxonomies to classify MOLP methods according to the stage of the method where the analyst needs to "extract" the preference information from the DM[5]. Four categories have been proposed concerning the stage in which the preference information (PI) is needed: (i) no articulation of PI, (ii) *a priori* articulation of PI, (iii) progressive articulation of PI, and (iv) *a posteriori* articulation of PI. With respect to the type of PI needed, this is in close relation with the four categories presented above.

As its name suggests, in no articulation of PI methods there is no need to access any kind of preference information from the DM. In fact, the methods compute the solution and the DM "only" needs to accept it. These methods have two main disadvantages. Firstly, the analysis needs to make assumptions regarding the DM preference. Secondly, since the DM is not involved in any way with the process of achieving the solution, (s)he may be very reluctant to accept that it is a good compromise solution. Given these two significant drawbacks, no method, in this category, will be addressed in this book.

For *a priori* methods, the information is either cardinal when the DM needs to provide judgement on a specific objective preference level (as in utility based methods which will not be addressed in this book[6]) or combine ordinal and cardinal information. Two of these methods will be studied: the lexicographic method in Chapter 8 and Goal Programming in Chapter 11.

In *a posteriori* methods, "all" non-dominated solutions are computed and the DM is asked to choose the most satisfactory solution (the compromise solution) among them. Since the decision moment is after the Pareto Front has been completed, no assumption or information is needed with respect to the

---

[5] Although this work is from 1980, it is still the most used taxonomy.

[6] For further information refer to Siskos et al. (2016).

DM's utility function. Two main limitations severely limit the application of these methods: 1) the large number of non-dominated solutions renders almost impossible for the DM to choose one that (s)he believes to be a good compromise among the objectives; 2) in real world problems, this is almost impossible to fully compute all non-dominated solutions (these methods are computationally very demanding). The ε-constraint method (presented in Chapter 8) and the weighted sum methods (Chapter 9) are able to fully determine the Pareto Front. However, the corresponding methodologies will not be discussed.

For methods that need the progressive articulation of PI (these are known nowadays as Interactive methods), the type of information needed by the DM is in the form of trade-offs. The methods based on explicit trade-offs subject the DM to answer several (sometimes very) hard questions. Methods that make use of implicit trade-off ask the DM to suggest acceptable achievement levels concerning the objective, which usually are easier to answer. Given the book length limitation, only the latter case will be covered (Chapter 10).

More formal introductions to MOLP main concepts can be found in literature. The interested reader is referred to Steuer (1986), Ehrgott (2005) or Wiecek et al. (2016) for more comprehensive presentations.

7.6 Exercises

7.6.1 Solved Exercises

Consider the following MOPL problem:

$$\text{Max } Z_1 = 2x + y$$
$$\text{Max } Z_2 = -x + y$$
$$\text{s.t.} \qquad x + 3y \geq 6$$
$$2x - y \leq 5$$
$$-x + y \leq 4$$
$$x + 2y \leq 12$$
$$x \leq 4$$
$$x, y \geq 0$$

(a) Solve independently each objective function presenting the optimal solution and the corresponding optimal value. Determine the efficient frontier.

Figure 7.9 shows the feasible region, set S. The optimal solution for Z_1 is vertex $(4, 4)$ to which corresponds the optimal value of $Z_1^* = 12$. Objective function Z_2 should be minimized. The gradient vector points towards the "decreasing direction", i.e., lines parallel to Z_2 drawn in the optimization direction (see the gradient vector) are decreasing in value. So, the optimal solution for Z_2 is vertex $(3, 1)$ and the optimal value is $Z_2^* = -2$.

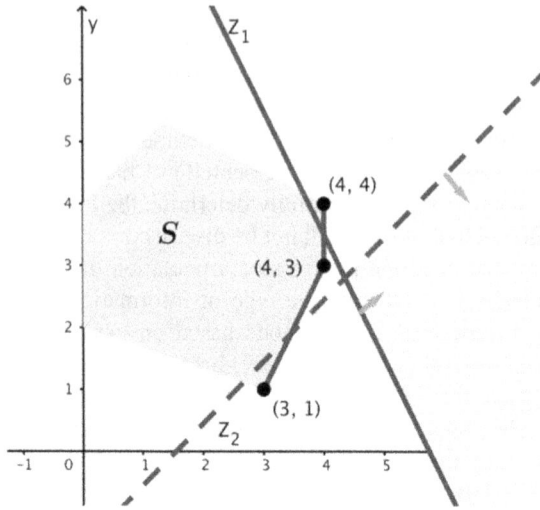

Figure 7.9: Feasible Region, Optimal Solutions, Dominance Cone and Efficient Frontier

The dominance cone (the light grey region where both gradient vectors are pointing to) shows that the efficient frontier is formed by two-line segments:

$$\lambda(3, 1) + (1 - \lambda)(4, 3) \text{ and } \lambda(4, 4) + (1 - \lambda)(4, 3) \text{ with } \lambda \in [0,1].$$

Any point in these segments are solutions that might be of interest to the Decision Maker.

(b) Graph the feasible region in the objectives space and highlight the solutions that might be of interest to the DM.

Vertex in S	Vertex in S_Z
(0, 2)	(2, 2)
(0, 4)	(4, 4)
(4/3, 16/3)	(8, 4)
(4, 4)	(12, 0)
(4, 3)	(11, -1)
(3, 1)	(7, -2)

Figure 7.10 shows the feasible region in the objectives space (set S_Z). The solutions that might be of interest to the DM are the ones forming the Pareto Front, i.e., any solution in the line segments

$$\lambda(12, 0) + (1 - \lambda)(11, -1) \text{ and } \lambda(11, -1) + (1 - \lambda)(7, -2) \text{ with } \lambda \in [0,1].$$

(c) Investigate the existence of weakly efficient solutions.
   This problem does not have weakly efficient solutions.
(d) The DM thinks he should choose the solution $(x, y) = (1, 5)$. Do you think this is a good solution?

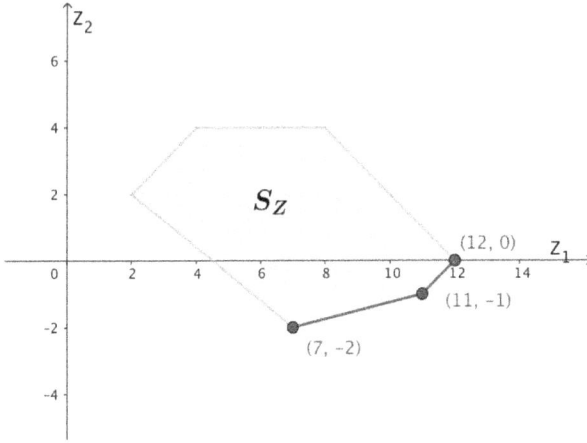

Figure 7.10: Feasible Region in the Objectives Space and Pareto Front

Solution $(1, 5)$ is a feasible solution, i.e., $(1, 5) \in S$. However, it is not an efficient solution (see Fig. 7.11). Thus, this is not a good solution for the DM since he can improve in both objectives if he chooses any of the efficient solutions.

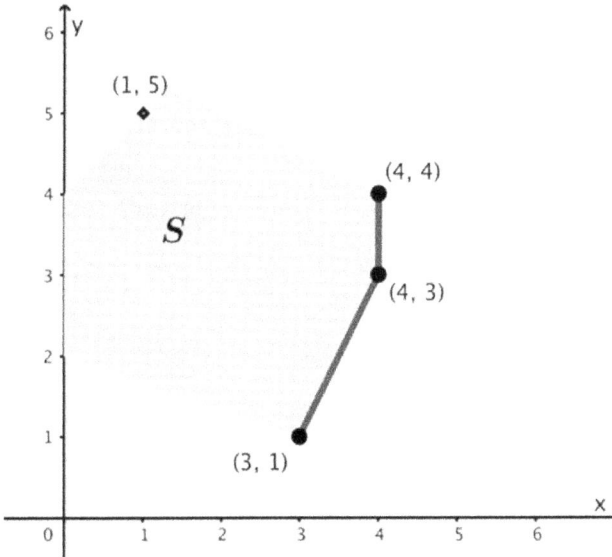

Figure 7.11: Solution $(1,5)$ in the Feasible Region

(e) Build the payoff table and determine the ideal and the nadir solutions.

Table 7.5 presents the payoff table. The nadir and the ideal solutions are, respectively, $Z^* = (12, -2)$ and $N^* = (7, 0)$.

Table 7.5: Payoff Table

	Z_1	Z_2
$x_1^* = (4, 4)$	12	0
$x_2^* = (3, 1)$	7	-2

(f) Adding a new objective function changes the Efficient Frontier?

The answer to this question is "it depends". Figure 7.12 (a) shows the case where the Efficient Frontier does not change. The new objective function Max $Z_3 = 3x - y$ does not change the dominance cone and consequently has not impact on the efficient frontier. Figure 7.12 (b) presents the case where the new objective function Min $Z_3 = 2x + 3y$ adds a new line segment to the Efficient Frontier.

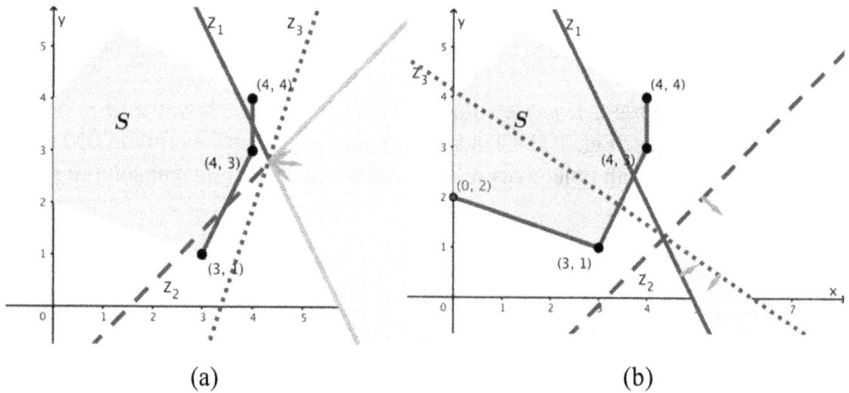

(a)   (b)

Figure 7.12: (a) Efficient Frontier Unchanged; (b) New Efficient Frontier

7.6.2   Proposed Exercises

1. Consider MOLP problem

$$\text{Max } Z_1 = y$$
$$\text{Max } Z_2 = x - 2y$$
$$\text{s.t.} \quad 5x - 2y \leq 40$$
$$3x + 5y \leq 55$$
$$-x + y \leq 3$$
$$x + y \geq 3$$
$$x, y \geq 0$$

(a) Solve independently each objective function presenting the optimal solution (or optimal solutions) and the corresponding optimal value.

(b) Graph problem on the objectives space and determine the set of non-dominated solutions.

2. Consider the MOLP problem

$$\text{Max } Z = (2x_1 + x_2; \, x_1 - x_2)$$
$$\text{s.t.} \qquad x_1 + x_2 \geq 2$$
$$-2x_1 + x_2 \geq 0$$
$$-x_1 + x_2 \leq 4$$
$$x_1, x_2 \geq 0$$

(a) Graph the feasible region in the decision space and solve each problem independently.
(b) Determine the efficient solution set.
(c) Investigate the existence of weakly efficient solutions.
(d) Graph the feasible region in the decision space and identify the Pareto Front.

3. Consider the MOLP problem

$$\text{Max } Z = (2x_1 + x_2; \, -x_1 + x_2)$$
$$\text{s.t.} \qquad x_1 + x_2 \geq 2$$
$$2x_1 - x_2 \leq 4$$
$$-x_1 + x_2 \leq 4$$
$$x_1 + 2x_2 \leq 12$$
$$x_1, x_2 \geq 0$$

(a) Graph the feasible region in the decision space and solve each problem independently.
(b) Determine the efficient solution set.
(c) Investigate the existence of weakly efficient solutions.
(d) Graph the feasible region in the objectives space and identify the Pareto Front.
(e) Do you agree with the statement "Solution $(x_1, x_2) = (1, 5)$ is of interest to the Decision Maker"? Justify your answer.
(f) Explain briefly what can happen to the results of question (b) if a new objective function is added to the problem.

4. Consider MOLP problem

$$\text{Max } Z_1 = -x + 3y$$
$$\text{Max } Z_2 = 3x + 3y$$
$$\text{Max } Z_3 = x + 2y$$
$$\text{s.t.} \qquad y \leq 4$$
$$x + 2y \leq 10$$
$$2x + y \leq 10$$
$$x, y \geq 0$$

(a) Graph the feasible region in the decision space and solve each problem independently.
(b) Determine the efficient solution set.
(c) Investigate the existence of weakly efficient solutions.
(d) Build the payoff table and determine the ideal and the nadir solutions.

5. Consider the MOLP problem

$$\text{Max } Z_1 = 2x + y$$
$$\text{Max } Z_2 = 0.5x + y$$
$$\text{s.t.} \qquad x \geq 3$$
$$y \geq 4$$
$$x \leq 8$$
$$-x + y \geq 3$$
$$0.5x + y \leq 12$$

(a) Graph the feasible region in the decision space and solve each problem independently.
(b) Determine the efficient solution set.
(c) Investigate the existence of weakly efficient solutions.
(d) Graph the feasible region in objectives space and identify the non-dominated solutions.
(e) Comment on the statement:
"Changing objective function Min $Z_2 = 0.5x + y$ to Max $Z_2 = -0.5x - y$ will have no impact on the Pareto Front."

6. Consider the following multi-objective linear programming problem whose feasible region S is represented in Fig. 7.13.

$$\text{Max } Z_1 = x - y$$
$$\text{Max } Z_2 = -2x + y$$
$$\text{s.t.} \quad (x,y) \in S$$

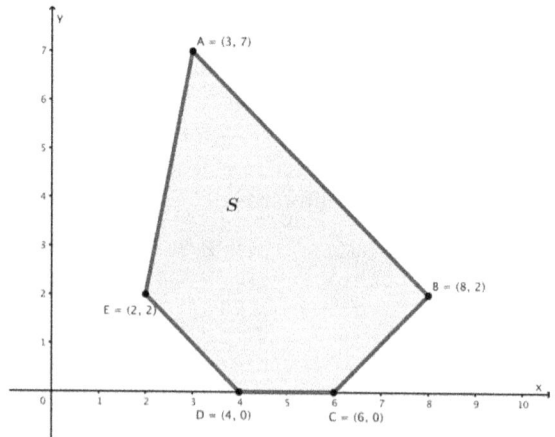

Figure 7.13: Feasible Region S.

(a) Put Y (yes) or N (no) indicating whether the line segment corresponds to solutions that may be of interest to the decision maker.

☐ *AB* ☐ *BC* ☐ *CD* ☐ *DE* ☐ *EA*

(b) Graph the region S in the objectives space, indicating the subset of solutions that may be of interest to the decision maker and the ideal solution.

(c) Build the payoff table and determine the ideal and the nadir solutions.

7. Consider the following multi-objective linear programming problem whose feasible region S is represented in Fig. 7.14.

$$\text{Max } Z_1 = -x + y$$
$$\text{Max } Z_2 = 4x - 2y$$
$$\text{Max } Z_3 = 4y$$
$$\text{s.t. } (x, y) \in S$$

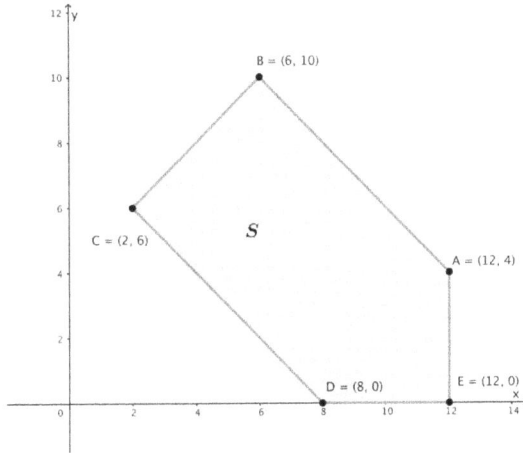

Figure 7.14: Feasible Region S.

(a) Draw the dominance cone and determine the efficient set.

(b) Investigate the existence of weakly feasible solutions.

(c) Build the payoff table and determine the ideal and the nadir solutions.

8. Jaime is preparing his study time for next week. He will have two tests (Algebra and Biochemistry). For each study hour of Algebra his girlfriend will offer him 5 dl of strawberry ice cream. But she already threatened him that she would reduce 2 dl of ice cream for every hour he studied Biochemistry. Note that his girlfriend is a Math student. Every hour of Biochemistry study will give him 4 units of satisfaction. But an hour of Algebra study will provide him with a unit of negative satisfaction. Jaime does not want to study more than 6 hours of Algebra, but the amount of Biochemistry study hours should not exceed in 3 hours the Algebra study period. He would like to maximize the amount of ice cream he will receive, but also the total satisfaction with the study.

This problem can be formulated as

$$\text{Max } Z_1 = 5x - 2y$$
$$\text{Max } Z_2 = -x + 4y$$
$$\text{s.t.} \qquad -x + y \le 3$$
$$x \le 6$$
$$x, y \ge 0$$

(a) Describe the meaning of the variables, objective functions and constraints.

(b) Graph the feasible region in the decision space and in the objectives space, marking the set of efficient solutions and the set of non-dominated solutions.

9. In a factory there are two types of machines used in the production lines of four products. The available capacities of each type of machine (in hours/week) and the number of processing hours/week required at each machine, per unit produced product, are presented in the table.

Machine	P1	P2	P3	P4	Availability
A	2	1	4	3	60
B	3	4	1	2	60

The company management wants to maximize profit, quality of P1, P2, P3 and P4 supply and the degree of employee satisfaction. Recent studies have made it possible to establish the unit contribution of each product to each objective.

	P1	P2	P3	P4
Profit	3	1	2	1
Quality	1	−1	2	4
Satisfaction	−1	5	1	2

(a) Formulate the problem as a multi-objective linear programming model.

(b) Using an optimization software (e.g. GAMS), determine a non-dominated solution that que optimizes:
   (i) the profit;
   (ii) the quality;
   (iii) the satisfaction.

10. Aunt Leonarda recently bought 22 kilograms of cregola, a very rare substance used in perfumery. She wants to use it to produce her two most famous perfumes, named as X and Y, for reasons of confidentiality. Aunt Leonarda wonders: "Each kiloliter of X gives me a profit of one thousand pilins, while with the perfume Y I manage to collect profits of 3 thousand pilins per kiloliter. But, each kiloliter of Y consumes two kilos of cregola, and with X the consumption is one to one..."

"Don't forget that to fulfil the production plan, at least three kiloliters of X and two of Y should be produced and that our distributor can only assure selling eight kiloliters of Y", added Blimunda, Aunt's secretary. "And there is something else, the total of X produced has to exceed the total production of Y by six kiloliters", Blimunda reminded her.

"What a mess, I want to minimize the consumption of cregola with this production, but I also want to make the highest profit.", admitted the friendly Aunt Leonarda.

Blimunda wrote something on paper and proudly presented the following formulation to Aunt Leonarda.

$$\text{Max } Z(-x_1 - x_2; x_1 + 3x_2)$$
$$\text{s.t.} \qquad x_1 \geq 3$$
$$x_2 \geq 2$$
$$x_2 \leq 8$$
$$x_1 - x_2 \leq 6$$
$$x_1 + 2x_2 \leq 21$$

(a) Comment on this formulation, establishing what is the relationship between each objective constraint/objective function with Aunt's and Blimunda's observations.

(b) Using the above formulation:

   (i) Graph the solution space and solve each objective independently.

   (ii) Identify the solutions that might be of interest to Aunt Leonarda (the efficient solutions) and investigate the presence of weakly efficient solutions.

   (iii) Graph the decision space and identify the Pareto Front.

References

Alves, M.J. and J.P. Costa (2009). An exact method for computing the nadir values in multiple objective linear programming. *European Journal of Operational Research*, 198(2), 637–646.

Anderluh, A., P.C. Nolz, V.C. Hemmelmayr and T.G. Crainic (2021). Multi-objective optimization of a two-echelon vehicle routing problem with vehicle synchronization and 'grey zone' customers arising in urban logistics. *European Journal of Operational Research*, 289(3), 940–958.

Antunes, C.H., M.J. Alves and J. Clímaco (2016). *Multiobjective Linear and Integer Programming*. Springer, New York, NY.

Ehrgott, M. (2005). *Multicriteria Optimization* (Vol. 491). Springer Science & Business Media.

Hwang, C.L., S.R. Paidy, K. Yoon and A.S.M. Masud (1980). Mathematical programming with multiple objectives: A tutorial. *Computers & Operations Research*, 7(1-2), 5–31.

Javanmard, M.E., S.F. Ghaderi and M.S. Sangari (2020). Integrating energy and water optimization in buildings using multi-objective mixed-integer linear programming. *Sustainable Cities and Society*, 62, 102409.

Javid, N., K. Khalili-Damghani, A. Makui and F. Abdi (2020). Multi-objective flexibility-complexity trade-off problem in batch production systems using fuzzy goal programming. *Expert Systems with Applications*, 148, 113266.

Kok, M. and J. Spronk (1985). A note on the pay-off matrix in multiple objective programming. *Omega*, 580–583.

Korhonen, P., S. Salo and R.E. Steuer (1997). A heuristic for estimating nadir criterion values in multiple objective linear programming. *Operations Research*, 45(5), 751–757.

Miettinen, K. (1998). *Nonlinear Multiobjective Optimization* (Vol. 12). Springer Science & Business Media.

Ribeiro, V.H.A. and G. Reynoso-Meza (2020). Ensemble learning by means of a multi-objective optimization design approach for dealing with imbalanced data sets. *Expert Systems with Applications*, 147, 113232.

Ruiz, A.B., R. Saborido, J.D. Bermúdez, M. Luque and E. Vercher (2020). Preference-based evolutionary multi-objective optimization for portfolio selection: A new credibilistic model under investor preferences. *Journal of Global Optimization*, 76(2), 295–315.

Siskos, Y., E. Grigoroudis and N.F. Matsatsinis (2016). UTA methods. *In:* Multiple Criteria Decision Analysis (pp. 315–362). Springer, New York, NY.

Steuer, R.E. (1986). Multiple Criteria Optimization: Theory, Computation and Applications. John Wiley and Sons, Inc.

Van Haveren, R., B.J. Heijmen and S. Breedveld (2020). Automatic configuration of the reference point method for fully automated multi-objective treatment planning applied to oropharyngeal cancer. *Medical Physics*, 47(4), 1499–1508.

Wiecek, M.M., M. Ehrgott and A. Engau (2016). Continuous multiobjective programming. *In: Multiple Criteria Decision Analysis* (pp. 739-815). Springer, New York, NY.

Winston, W.L. (2002). Introduction to Mathematical Programming: Applications and Algorithms, Duxbury.

Zeleny, M. (1998). Multiple criteria decision making: Eight concepts of optimality. *Human Systems Management*, 17(2), 97–107.

Lexicographic and ε-Constraint Methods

8.1 Introduction

The Lexicographic and the ε-constraint methods are two of the most well-known methods in Multi-Objective Linear Programming (MOLP). Both methods fit in to the category of Reduced Feasible Region Methods, which then falls within the category of "a priori" preference aggregation methods within a broader classification system for MOLP methods (as presented in Chapter 7). In short, both methods transform the MOLP problem into a single objective problem by optimizing one objective function and taking the remaining objectives (all at once, or step by step) as constraints. The Lexicographic method asks for the Decision Maker (DM) to sort all objective functions from the most important to least important one. Then, it takes one objective function at a time and finds its optimal solution within a reduced feasible region. This reduction is made considering the previous solved objectives and setting them greater or equal to some aspirations level proposed by the DM. The ε-constraint method asks the DM for the most important objective and solves the single objective problem considering only this function, while all the remaining objective functions are set as constraints. The right-hand side of these new constraints are achievement levels suggested by the DM for each objective.

Formally, a MOLP problem can be formulated as follows:

$$\text{Max } (Z_1(x), Z_2(x), \dots, Z_k(x))$$

$$\text{s.t.} \qquad x \in S$$

with x the vector of decision variables, $Z_1(x)$, $Z_2(x)$, ..., $Z_k(x)$ the k objective functions and S the feasible region defined by the model constraints[1].

The Example 8.1 presented below will serve as a baseline problem throughout this chapter, allowing for a better understanding of each method's main features. In all examples presented in this chapter's main body, all objective functions

---

[1] For brevity, some fundamental concepts addressed in Chapter 7 will not be defined in this chapter. However, a supplementary chapter is available for free with all the main definitions and concepts needed to fully understand the topics addressed in this chapter.

will be maximized. At the end one example having one objective function to be minimized will be solved by both methods.

Example 8.1

Suppose the DM aims at determining the compromise solution of the three objective functions problem presented below. For the integrality of this chapter, Fig. 8.1 presents the feasible region (set S), the three objective functions and their gradient vectors. The optimal solutions are also presented if each objective function is optimized individually[2].

Max $Z_1 = 20x + 10y$

Max $Z_2 = -80x + 40y$

Max $Z_3 = y$

s.t. $x + y \geq 20$

$2x - y \leq 40$

$-x + y \leq 80$

$x + y \leq 80$

$x, y \geq 0$

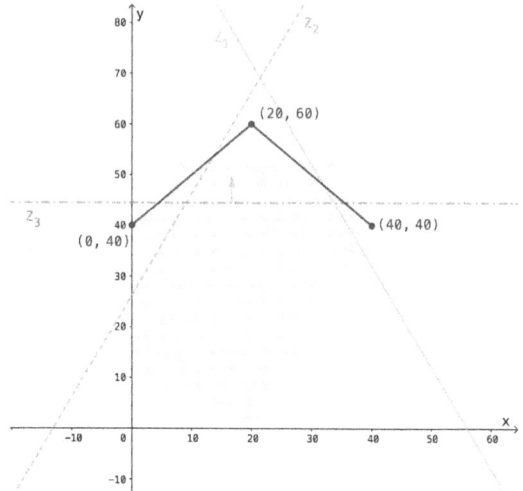

Figure 8.1: Feasible Region (S), Objective Functions and their Optimal Solutions, and Efficient Frontier.

By solving independently each objective function, one determines that the optimal value of the first objective function is $Z_1^* = 1200$ attained at vertex $x_1^* = (40, 40)$, the extreme point $x_2^* = (0, 40)$ is the optimal solution of Z_2 corresponding to the optimal value of $Z_2^* = 1600$, and $x_3^* = (20, 60)$ is the optimal solution for the last objective function with an optimal value of $Z_3^* = 60$.

The set of efficient solutions are the line segments $[x_2^* \, x_3^*]$ and $[x_3^* \, x_1^*]$.

This chapter is organized into three parts. The first one addresses the lexicographic method, presented using a motivational example. The algorithm is then formally defined (section 8.2.1). The capability of identifying weakly efficient solutions is one of the method's main features (section 8.2.2). Some final remarks will be drawn in the last section of the method (section 8.2.3). The second part of this chapter focus on the ε-constraint method, one of the most popular methods for Pareto Front approximation. This section starts with the method's

---

[2] For further details on how to graphically solve a linear programming problem, the interested reader should refer to any Operations Research reference book as, for example, Winston (2002).

formal presentation. Sections 8.3.1 and 8.3.2 will address the methods' main limitations: infeasibility (often the choice of achievement levels will render the model with an empty feasible region) and weakly efficiency (the method is unable to detect whether a compromise solution is weakly efficient). Then, before the last remarks (section 8.3.3), the AUGMENCON method is presented. This is a very powerful method to approximate the Pareto Front while avoiding weakly efficient solutions and minimizing the number of impossible problems (Mavrotas 2009). This chapter ends with a set of exercises addressing the topics covered.

8.2 The Lexicographic Method

The lexicographic method is a very simple method for determining a compromise solution in MOLP. It asks the DM to sort out all the objective functions and to evaluate, step by step, how much they are willing to compromise on the optimal value of the objective that has just been solved.

In this method the first step is then to ask the DM which is the most important of the objectives, which is the second most important, and so on until the last objective is reached. In the end, the DM will have ranked all objectives in decreasing order of preference (no ties are allowed). Then the first objective function (the most important one) is solved as a single linear problem subjected to the problem original feasible region (set S). Next, the second (most important) objective is solved again as a single-objective problem but now the feasible region has a new constraint. This new constraint sets a lower bound on the first objective function (the one solved previously), so that it does not fall below the values that the DM is not willing to concede on. Consequently, when searching for the optimal solution of the second most important objective, at least a minimum achievement level is assured for the first objective function. Having determined the (second) optimal solution, the third most important objective function is solved, again as a single-objective problem and again having the (previous) feasible region reduced by a new constraint. As stated previously, this constraint will limit the second objective function. Note that the first added constraint is not to be removed. This procedure continues until all objective functions have been evaluated or until the feasible region has been reduced to a single solution. At most, one will need to perform k steps (as many as the number of objective functions).

What should be the values to impose in each constraint? These values should be provided by the decision maker (DM), since they depend on the DM's preferences. These are the minimum achievement levels the DM wishes to attain.

Example 8.2

Taking the Example 8.1 problem, let's assume the DM considers Z_3 is the most important objective function followed Z_1 and finally Z_2. The objective functions in decreasing order are

$$Z_3 \gg Z_1 \gg Z_2$$

where "≫" reads as "more important than".

As mentioned above, the first step of the method is to solve

$$\text{Max } Z_3$$
$$\text{s.t. } x \in S.$$

The optimal solution[3] is vertex $x_3^* = (20, 60)$ with an optimal value of $Z_3^* = 60$.

In the second step, a new single-objective problem must be solved considering the second most important objective, which is Z_1. The model below formulates the problem to be solved at the second iteration.

$$\text{Max } Z_1 = 20x + 10y$$
$$\text{s.t.} \quad x + y \geq 20$$
$$2x - y \leq 40$$
$$-x + y \leq 80$$
$$x + y \leq 80$$
$$y \geq 60 - \varepsilon_3$$
$$x, y \geq 0$$

The new constraint is $Z_3 \geq Z_3^* - \varepsilon_3$, where ε_3 value is as large or as small as the DM wishes. Note that, if $\varepsilon_1 = 0$ the new feasible region (S_1) will be reduced to vertex $(20, 60)$, since the added constraint will be $y \geq 60$. This will make the vertex $(20, 60)$ again the optimal solution for the second step. In fact, when adding a constraint limiting the objective function to take its optimal value, the lexicographic approach ends unless that objective function has multiple optimal solutions (this will be explored later in the chapter). So, in order to take the second most important objective into account, the DM needs to surrender part of the optimal value of objective Z_3. Suppose

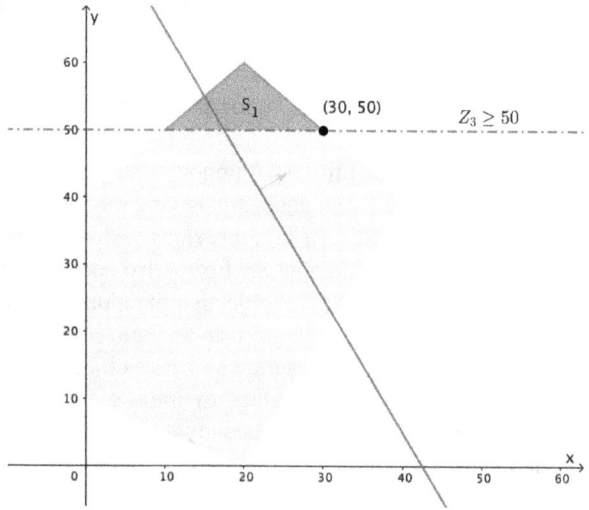

Figure 8.2: Feasible Region (S_1) and Optimal Solution (vertex (30,50)) for the Lexicographic Step 2

---

[3] This problem has been solved in Example 8.1.

the DM decides to let go of 10 units. This will reduce the feasible region S to S_1 (Fig. 8.2). The optimal solution for the second most important objective function, Z_1, is vertex (30, 50) with Z_3^* (30, 50) = 1100.

Since the problem has three objective functions, the last function Z_2 will now be optimized. The model formulation is now:

Max $Z_2 = -80x + 40y$

s.t. $x + y \geq 20$

$2x - y \leq 40$

$-x + y \leq 80$

$x + y \leq 80$

$y \geq 60 - \varepsilon_3$

$20x + 10y \geq 1100 - \varepsilon_1$

$x, y \geq 0$

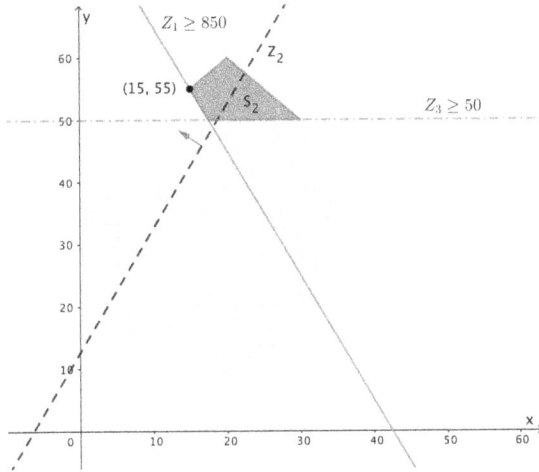

Figure 8.3: Feasible Region (S_2) and Optimal Solution (Vertex (15, 55)) for Lexicographic Step 3.

Suppose the DM accepts an $\varepsilon_1 = 250$. This will further reduce the feasible region into polyhedron S_2 (Fig. 8.3). The Z_2 optimal solution in S_2 is vertex (15, 55) corresponding to the optimal value $Z_2^* = 1000$.

Since this is the third objective function, the lexicographic method has reached its final step. Therefore, the DM's compromise solution has been found and is

$$(x, y) = (15, 50) \text{ with } Z_1 = 850, Z_2 = 1000 \text{ and } Z_3 = 55.$$

8.2.1 The Algorithm

Generically, the lexicographic method has the following steps:

1. Rank all objectives from the most important to the least important, ties are not allowed. Suppose, without loss of generality that, $Z_1 \gg Z_2 \gg ... \gg Z_k$, where "$\gg$" reads as "more important than".
2. Solve the problem

$$\text{Max } Z_1(x) \text{ s.t. } x \in S_1,$$

where S_1 is the feasible region of the original problem and x is the n-dimensional vector of the decision variables.

3. For $i = 2, \ldots, k$ solve

$$\text{Max } Z_i(x) \text{ s.t. } x \in S_i,$$

where $S_i = S_{i-1} \cup \{Z_{i-1} \geq z^*_{i-1} - \varepsilon_{i-1}\}$ and z^*_{i-1} the optimal value of the objective function optimized in the previous iteration, and ε_i is the maximum value the DM is willing to reduce from the optimal value of objective function i. Notice the ε_i values should be non-negative. If negative they will render the corresponding problem infeasible.

4. The compromise solution is the optimal solution of

$$\text{Max } Z_k(x) \text{ s.t. } x \in S_k.$$

8.2.2 The Lexicographic Method as a Tool to Identify Weakly Efficient Solutions

In a n-dimensional variable space, the great majority of optimization software is not capable of determining if some objective function has alternative optimal solutions. Most algorithms will stop after identifying the first optimal solution, since computing all alternative optimal solutions can be computationally harder than solving the original problem. The lexicographic approach is a very useful tool to identify objective functions in MOLP that have multiple optimal solutions.

Example 8.3

Example 8.1 will be slightly adapted to accommodate the existence of multiple optimal solutions. Take a new objective function $Z_1' = 10x + 10y$. Its optimal solutions are all points within the line segment (Fig. 8.4)

$$\lambda(20, 60) + (1 - \lambda)(40, 40), \lambda \in [0, 1].$$

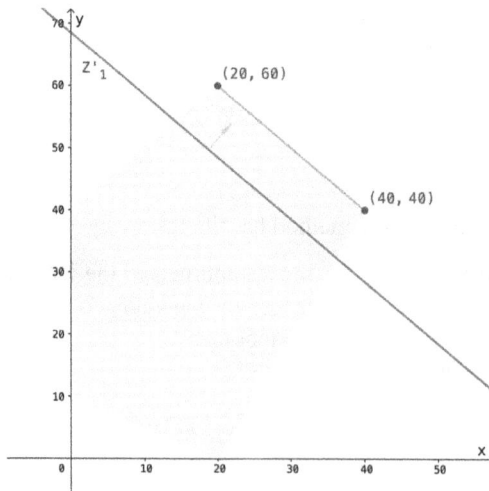

Figure 8.4: Optimal Solutions for Z_1'

Take objective functions Z_2 and Z_3 and solve each one of them independently with the additional constraint

$$Z_1' = 800 \Leftrightarrow 10x + 10y = 800.$$

For both objective functions, the optimal solution is vertex (20, 60). This means that all other solutions on the line segment $\lambda(20, 60) + (1 - \lambda) (40, 40)$, $\lambda \in [0, 1]$, will decrease the value of Z_2 and Z_3 without improving the value of Z_1'. It can then be concluded that they all are weakly efficient solutions and therefore they are no longer good compromise solutions. The Pareto Front is then

$$\lambda(0, 40) + (1 - \lambda) (20, 60), \lambda \in [0, 1].$$

8.2.3 Final Remarks

The Lexicographic method should only be applied when the DM is able to rank without ties all the objective functions. In other words, when the DM is confident enough to compare all objective functions and is able to assess their relative importance, two objective functions are in the same position. If this is not the case, a different method should be applied. For instance, the ε-constraint method that will be presented next within this chapter. For further details on the theoretical background of the Lexicographic method the reader is referred to Bouyssou et al. (2006).

In this method, DM's preferences are modelled in two different moments: when sorting the objective functions according to their importance or significance, and when setting the ε_i values.

The main advantages of the method are:

- It does not require that the objective functions be normalized as other methods do (e.g. the weighted sum method that will be studied in Chapter 9),
- It just requires an ordinal ranking of the objective functions, and
- Since the method alters the original feasible region it can produce efficient solutions other than the vertices of the original problem.

While the main disadvantages are:

- It may entail the solution of many single-objective problems to obtain just one compromise solution, and
- Since additional constraints are imposed, it requires the setting of the lower bounds which can be difficult to assess by the DM.

8.3 ε-Constraint Method

The ε-constraint method is also a reduced feasible region method and was first published by Haimes et al. (1971). It converts $k - 1$ objectives into constraints, leaving to be optimized the objective considered by the DM as the most important. It then solves the single objective problem considering only this function, while all the remaining objectives are constraints bounded by values also proposed by the DM. The bounds are largely determined by the DM's experience. The main idea is to set a model that produces a good compromise solution.

Let's assume, without the loss of generality that Z_1 is the most important objective function amongst the k objective function comprising the MOLP problem. Formally the ε-constraint method can be defined as follows:

$$\text{Max } Z_1(x)$$
$$\text{s.t. } Z_2(x) \geq \varepsilon_2$$
$$\dots$$
$$Z_k(x) \geq \varepsilon_k$$
$$x \in S.$$

By varying the constraints limits (ε_i, $i = 2, \dots, k$) one reaches efficient solutions, which means reaching different compromise solutions that may be of interest (or not) to the DM. However, as presented above this method is not able to detect when solutions are weakly efficient, but let's leave this issue for a bit later in this chapter. For now, let's exemplify the method taking the two first objective functions of Example 8.1.

Example 8.4

Suppose the DM considers Z_1 as the most important objective. The ε-constraint model to be solved is given below where ε is a value suggested by the DM. As for the lexicographic method, setting "good" ε values may be very challenging for the DM. The payoff table is a helpful tool since it can help the DM to set adequate ε values (Table 8.1).

$$\text{Max } Z_1 = 20x + 10y$$

s.t.
$x + y \geq 20$
$2x - y \leq 40$
$-x + y \leq 80$
$x + y \leq 80$
$-80x + 40y \geq \varepsilon$
$x, y \geq 0$

Table 8.1: Payoff Table of Example 8.4

	Z_1	Z_2
$x_1^* = (40, 40)$	1200	−1600
$x_2^* = (0, 40)$	400	1600

The DM can set any value of $\varepsilon \in [-1600, 1600]$. Assuming $\varepsilon = 1000$, Figure 8.5 (a) shows the reduced feasible region and the corresponding optimal solution. With such ε for Z_2, the compromise solution is $x^* = (15, 55)$ with $Z_1^* = 850$ and $Z_2^* = 1000$.

Different ε values will return different compromise solutions as shown in Figure 8.5 (b) and (c), where ε has been set to 1200 and 0, respectively. Notice that with $\varepsilon \in [-1600, 1600]$ if it is set to its extreme values, the feasible region will either be reduced to a single point (when setting it to the maximum value) or will be the original feasible region if it is set to the minimum value (this objective function is parallel to constraint $2x - y \leq 40$).

When the MOLP has 3 or more objective functions, setting adequate ε values can be considerably more challenging.

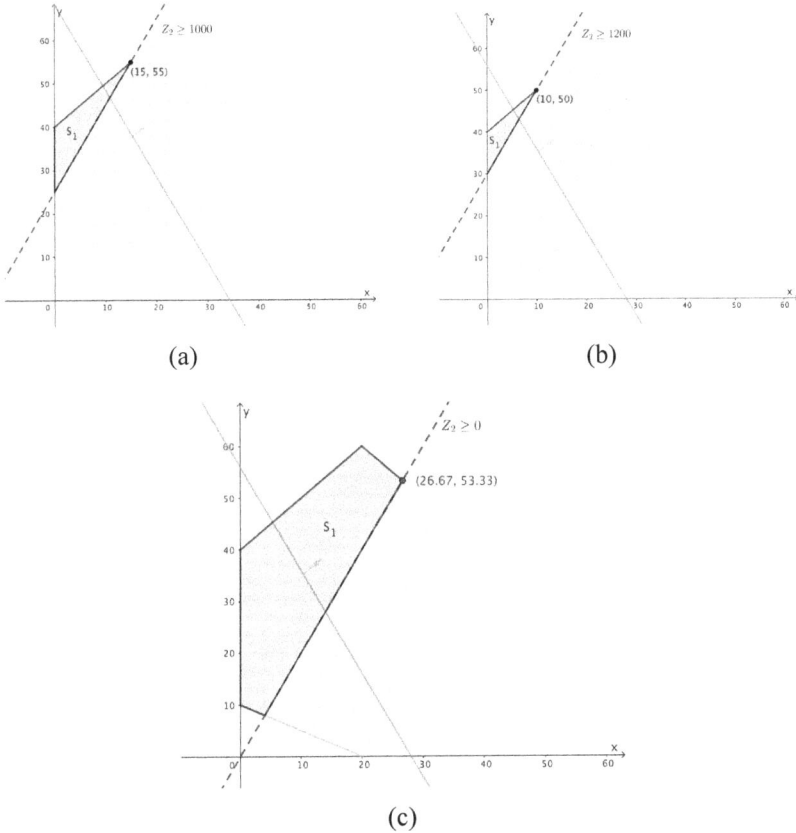

(a)

(b)

(c)

Figure 8.5: Example 8.4 Reduced Feasible Region with
(a) $\varepsilon = 1000$, (b) $\varepsilon = 1200$ and (c) $\varepsilon = 0$.

Example 8.5

Let's now take Example 8.1 with its three objectives. The optimal solutions and optimal values of the objective functions are already known. The payoff table is presented in Table 8.2.

Table 8.2: Payoff Table of Example 8.1.

	Z_1	Z_2	Z_3
$x_1^* = (40, 40)$	1200	−1600	40
$x_2^* = (0, 40)$	400	1600	40
$x_3^* = (20, 60)$	1000	800	60

The ideal solution is $Z^* = (1200, 1600, 60)$ and the nadir solution $N^* = (400, -1600, 40)$. Interesting values for ε_i are at most $\varepsilon_1 \in [400, 1200]$, $\varepsilon_2 \in [-1600, 1600]$ and $\varepsilon_3 \in [40, 60]$.

Let's assume Z_3 is the most important objective and let's now suppose the DM chooses $\varepsilon_1 = 600$ and $\varepsilon_2 = 920$. The corresponding ε-constraint model is given below.

The feasible region is reduced to the S_1 (Fig. 8.6). The Z_3 optimal solution in S_1, the compromise solution, is vertex $(17, 57)$ with $Z_3(17, 57) = 57$, $Z_1(17, 57) = 910$ and $Z_2(17, 57) = 920$.

Being the Z_2 constraint one of the active constraints[4], different ε_2 values lead to different compromise solutions. In contrast, the Z_1 constraint is not active[5] and therefore, different values of ε_1 will keep the optimal solution unchanged (e.g., all values below 910; with $\varepsilon_1 = 910$, Z_1 becomes an active constraint).

$$\text{Max } Z_3 = y$$

$$\text{s.t.} \quad x + y \geq 20$$
$$2x - y \leq 40$$
$$-x + y \leq 80$$
$$x + y \leq 80$$
$$20x + 10y \geq 600$$
$$-80x + 40y \geq 920$$
$$x, y \geq 0$$

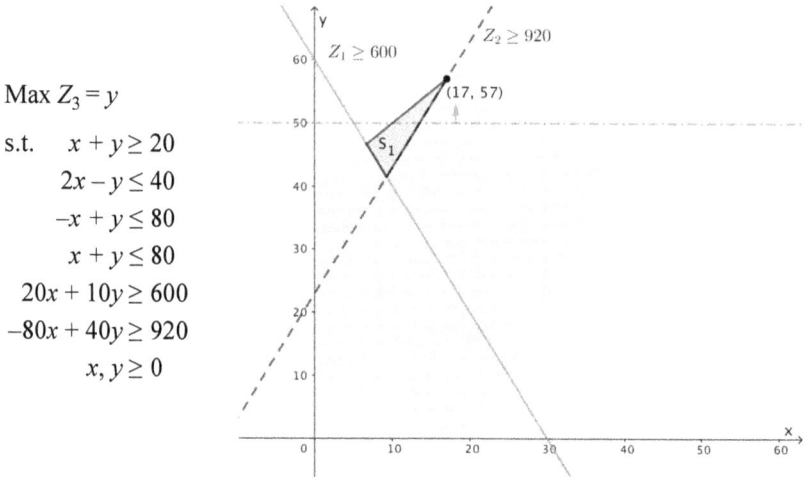

Figure 8.6: Reduced Feasible Region with $\varepsilon_1 = 600$ and $\varepsilon_2 = 920$ and Corresponding Optimal Solution

8.3.1   Infeasibility

In Example 8.5 if $\varepsilon_1 > 910$ and $\varepsilon_2 = 920$, the problem becomes infeasible since S_1 is empty. This is one of the limitations of ε-constraint method. When tackling three or more objective functions, the ε's values can easily be such that the feasible region becomes empty and consequently the problem is infeasible.

---

[4] Active constraints are the constraints forming the optimal vertex. In other words, an active constraint is a constraint that is limiting the objective function value.
[5] It is said to be inactive.

Example 8.6

Suppose the DM is willing to set the aspiration levels for the objectives to be at most 10% of the optimal values. Thus, it translates into $\varepsilon_1 = 1080$ and $\varepsilon_2 = 1440$. Such limits render the problem infeasible (Fig. 8.7 (a)). In order to proceced, the DM should revise the suggested limits until a feasible solution is found. Suppose, now, the decision maker wished Z_1 value to be at least 15% of its optimal value. This means $\varepsilon_1 \geq 1020$. At its lowest value, the problem is feasible if $\varepsilon_1 \leq 560$ (Fig. 8.7 (b)). Therefore, Z_2 needs to no larger than 35% of its optimal value so that a compromise solution can be found[6].

Although not an iterative approach, the ε-constraint method is very often used as such, since whenever a new compromise solution is computed, the DM can evaluate its adequacy and propose a new set of ε values if (s)he finds it "not good enough". At some point the DM will be satisfied with a solution (or tired of the process) and it will stop.

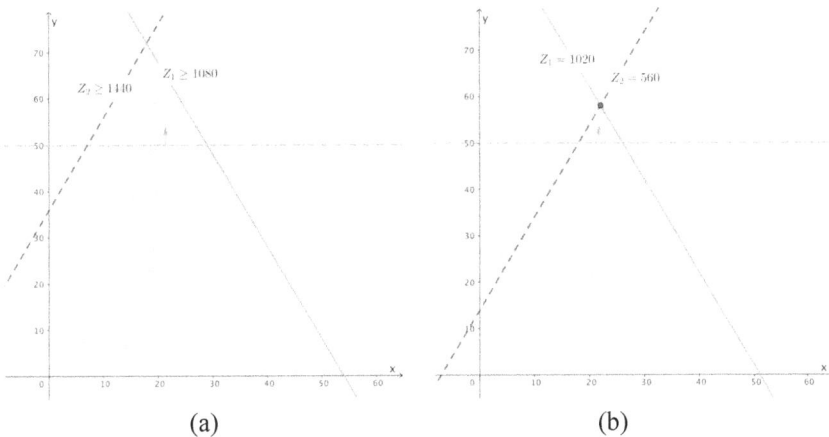

(a)   (b)

Figure 8.7: ε_i Values Leading to (a) Infeasible Problem and (b) a Single Feasible Solution.

8.3.2 Weakly Efficiency

Another limitation of the ε-constraint method is its inability to detect weakly efficient solutions. These solutions are not good solutions for a rational DM since they are dominated by other efficient solutions (see Chapter 7 for more details about weakly efficiency). This limitation can be overcome in the objective function of the ε-constraint method if it is replaced by

$$\text{Max } Z_1(x) + \sum_{p=2}^{k} \delta_p \cdot Z_p(x) \qquad [1]$$

---

[6] To compute the upper bound for ε_2, one should solve the system $20x + 10y = 1020$, $x + y = 80$, $-80x + 40y = \varepsilon_2$.

with δ_p a small positive value (Antunes et al. 2016). The second term "pushes" the optimal solutions to reach the highest possible values for the remaining objectives.

Example 8.7

Let's take Example 8.3. When asked about the most important objective, the DM was very clear it was objective Z_1'. She also wanted objectives Z_2 and Z_3 to be at least –500 and 11, respectively. The (original) ε-constraint model is formulated below. Figure 8.8 shows the reduced feasible region of this model.

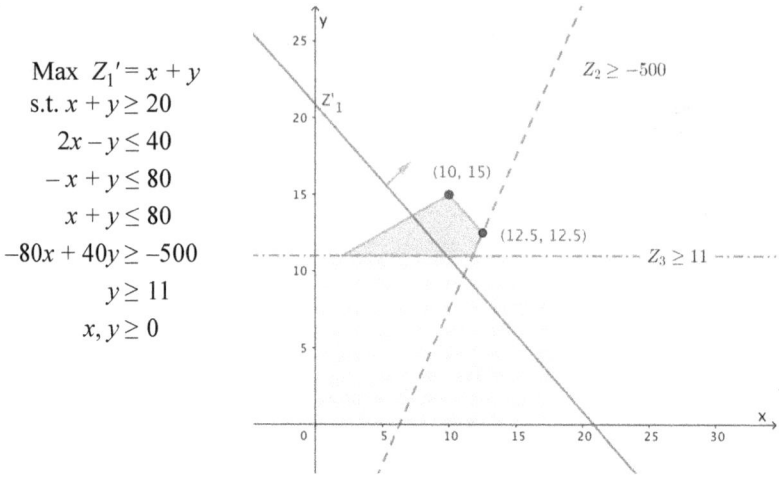

$$\text{Max } Z_1' = x + y$$
$$\text{s.t. } x + y \geq 20$$
$$2x - y \leq 40$$
$$-x + y \leq 80$$
$$x + y \leq 80$$
$$-80x + 40y \geq -500$$
$$y \geq 11$$
$$x, y \geq 0$$

Figure 8.8: Reduced Feasible Region of Example 8.7.

Note that any solution lying in segment $\lambda(10, 15) + (1 - \lambda)(12.5, 12.5)$, $\lambda \in [0, 1]$ is optimal for Z_1'. However, only vertex $(10, 15)$ is an efficient solution (all other solutions are weakly efficient). This issue is captured by the new objective function given in [1]. For this example and considering $\delta_2 = 10^{-6}$ and $\delta_2 = 10^{-2}$, this objective function is

$$\text{Max } Z_1' + 10^{-6} Z_2 + 10^{-2} Z_3. \qquad [2]$$

Table 8.3 shows the values for the three objective functions and objective function [2]. One can see that although both vertices present the same value for Z_1', the values differ for Z_2 and Z_3, and this difference is captured by the new objective function.

Table 8.3: Objective Functions Values

	Z_1'	Z_2	Z_3	[2]
(10, 15)	25	–200	15	25.1498
(12.5, 12.5)	25	–500	12.5	25.1245

8.3.3 How to Approximate the Pareto Front using ε-Constraint Method?

Mavrotas (2009) proposed an improved version of the ε-constraint method, the AUGMECON. This algorithm is an iterative procedure that computes several efficient solutions which allows for the approximation of the Pareto Front. It only produces Pareto optimal solutions both for the payoff table as well as during the solution generation process. It also has a better computational time for problems with 3 or more objective functions than the "straightforward" ε-constraint method since it is able to detect if a set of ε values lead to infeasible problems and, if so, it moves the process to the "next" non-empty feasible region.

The AUGMECON model is given below and assumes without loss of generality that Z_1 is the most important objective for the DM.

$$\text{Max } Z_1(x) + \delta(s_2/r_2 + s_3/r_3 + \ldots + s_k/r_k)$$
$$\text{s.t.} \quad Z_2(x) - s_2 = \varepsilon_2$$
$$Z_3(x) - s_3 = \varepsilon_3$$
$$\ldots$$
$$Z_k(x) - s_k = \varepsilon_k$$
$$x \in S, s_i \geq 0$$

where $\varepsilon_2, \varepsilon_3, \ldots, \varepsilon_k$ are the bounds for each objective in each algorithm iteration, $s_2, s_3, \ldots, s_k$ are the surplus variables[7] for each constraint, and $r_2, r_3, \ldots, r_k$ are the ranges of the corresponding objective functions. The range values are calculated from the payoff table and their role is to remove any scaling issues[8]. δ should be "small enough not to disturb" the optimal solution, $\delta \in [10^{-6}, 10^{-3}]$. Such a small value will force the model to produce only efficient solutions, avoiding the generation of weakly efficient solutions. The ranges r_i are computed from the payoff table.

In short, the AUGMECON algorithm[9] performs as follows:

1. Build the payoff table applying the lexicographic method for each function
2. Get the range of each one of the $k - 1$ objective functions that are going to be used as constraints from the payoff table
3. Divide the range of each objective function into equal sub-intervals (say q_i for Z_i). Thus, $q_i + 1$ equidistant points will be defined. In doing so, a total of $q_i + 1$ grid points have been defined and will be used as ε_i values.
4. Solve the model for each of the grid points, starting at the lowest point in the grid (when ε_i values are the nadir point coordinates). If the model is feasible,

---

[7] For more details on "surplus" variables, refer to Winston (2002).
[8] Scaling issues will be addressed in depth in Chapter 9.
[9] This algorithm is available in the GAMS library at https://www.gams.com/31/gamslib_ml/libhtml/gamslib_epscm.html

register the efficient solution and move to the next ε_i by gradually increasing the value by q_i. If the model is infeasible, move to the next objective function.

Notice that by starting at the grid point with the lowest ε_i, the algorithm starts at the most relaxed version of the model (the "largest" feasible region). By gradually increasing the ε_i values (rendering the feasible region more constrained), one assures that the minimum number of infeasible problems will be solved. With this process, whenever an infeasibility is reached one can stop increasing the ε value of the corresponding objective function since a larger ε value will also render the problem infeasible. The algorithm will run at most $(q_2 + 1) \cdot (q_3 + 1) \dots (q_k + 1)$ times.

The number of grid points depends, not only, on the density one desires for the Pareto Front but also on the computational times. The larger the number of grid points, the higher the computational time. Therefore, a tradeoff exists between the density of the Pareto Front and its computational time.

AUGMECON is presented with detail in the flowchart depicted in Fig. 8.9.

Example 8.8

The example[10] below is a MOLP with three objective functions (on the left) and the corresponding ε-constraint model (on the right) assuming Z_1 as the most important objective function.

$$\text{Max } Z_1 = x_1 \qquad\qquad \text{Max } Z_1 = x_1$$
$$\text{Max } Z_2 = x_2 \qquad\qquad \text{s.t.} \quad x_2 \geq \varepsilon_2$$
$$\text{Max } Z_3 = x_3 \qquad\qquad\qquad x_3 \geq \varepsilon_3$$
$$\text{s.t. } x_1 + x_2 + x_2 \leq 1 \qquad x_1 + x_2 + x_2 \leq 1$$
$$x_1, x_2, x_3 \geq 0 \qquad\qquad x_1, x_2, x_3 \geq 0$$

Figure 8.10 depicts the feasible region in the objectives space and some steps of the AUGMECON method. The range of each one of the two objective functions is divided in 10 intervals (11 grid points). In the course of the method, when $\varepsilon_2 = 0.5$ and $\varepsilon_3 = 0.6$ the problem becomes infeasible (since $x_2 + x_2 \leq 1$). In this case there is no need to check for $\varepsilon_3 = 0.7$, $\varepsilon_3 = 0.8$, etc. (marked with a star in Fig. 8.10). So, the algorithm exits from the ε_3 loop and directly goes to $\varepsilon_2 = 0.6$ and $\varepsilon_3 = 0$.

8.3.4   Final Remarks

The ε-constraint method is very common in MOLP literature both for determining a compromise solution as for the approximation of the Pareto Front. In this method, DM's preferences are modelled in two different moments: when the most important objective function is chosen, and when setting the ε_i values.

---
[10] Example proposed by Mavrotas (2009).

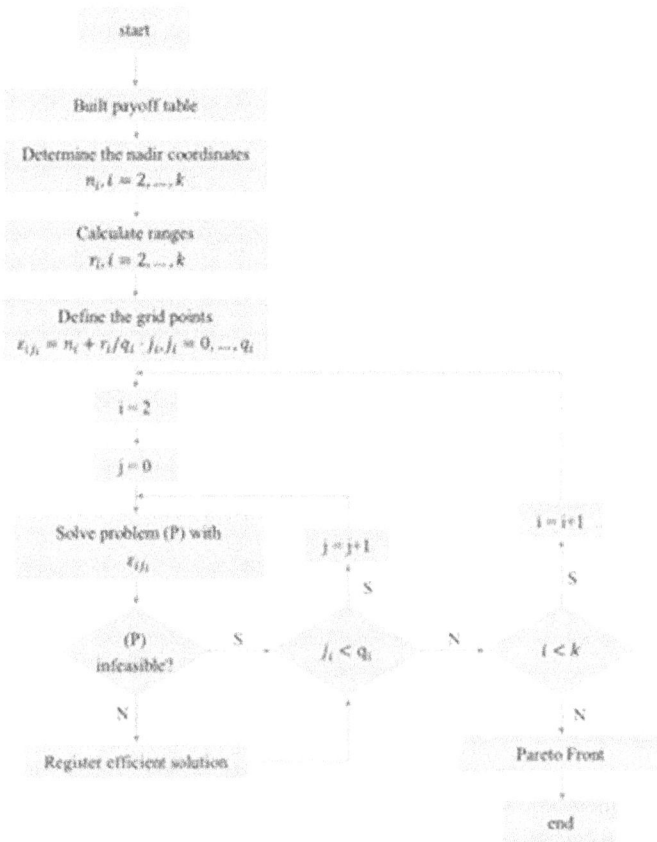

Figure 8.9. AUGMECON algorithm.

The main advantages of the method are:

- It does not require the objective functions be normalized as other methods do (e.g. the weighted sum method that will be studied in chapter 9),
- It only requires the choice of one of the objective functions instead of a full rank as the lexicographic method,
- Since the method alters the original feasible region it can produce efficient solutions other than the vertices of the original problem, and
- It can easily be adapted to provide only Pareto optimal solutions.

While the main disadvantages of the ε-constraint method are:

- It can easily lead to unfeasible problems when dealing with three or more objective functions,
- May propose weakly efficient solutions which are not "good" compromise solution for a rational DM, and

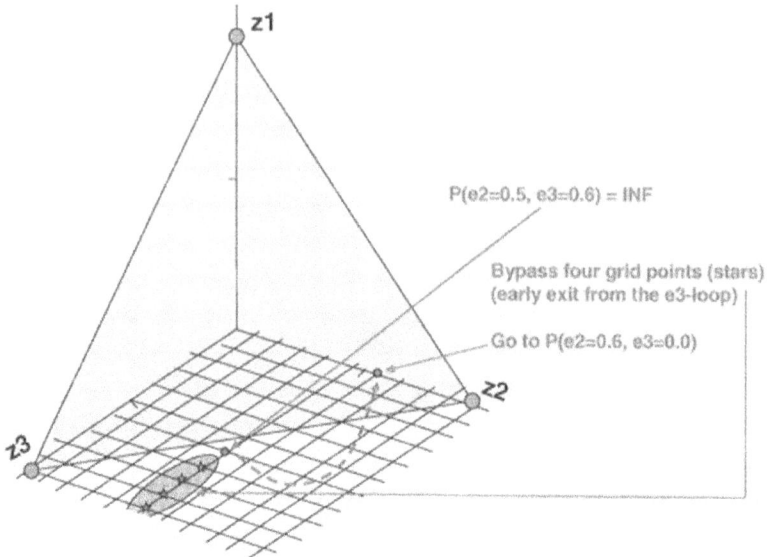

Figure 8.10: Graphical Representation of the Pareto Front and the AUGMECON's
Feature of the Early Exit from the Nested Loops when Detecting Infeasibility
(from Mavrota, 2009)

- Since additional constraints are imposed, it requires the setting of the lower
  bounds which can be difficult to be evaluated by the DM.

The AUGMECON method is presented in this chapter as a tool to overcome
some of the method's limitations. Other methods can be found in the literature
where authors aim at improving the ε-constraint method's performance but they
are out of the scope of this book. The interested reader may check the works of
Laumanns et al. (2006), Ehrgott and Ruzika (2008), Mavrotas and Florios (2013),
to name a few. For other Pareto Front approximation methods, refer to Ruzika and
Wiecke (2005) or Herzel et al. (2021).

8.4   Exercises

8.4.1   Solved Exercises

Consider the MOPL problem with three objective functions presented below.
Figure 8.11 shows the feasible region (set S), the objective functions and the
corresponding gradient vectors and optimal solutions. It also depicts the Efficient
Frontier. Suppose the DM's preferences concerning the objectives are $Z_3 \gg Z_2 \gg Z_1$ and then they would not like to compromise more than 20% in each objective
function value.

(a) Determine the compromise solution that reflects the DM's preferences
    according to the lexicographic method.

$$\text{Max } Z_1 = 2x + y$$
$$\text{Min } Z_2 = -x + y$$
$$\text{Max } Z_3 = 3x - y$$
$$\text{s.t. } x + 3y \geq 6$$
$$2x - y \leq 5$$
$$-x + y \leq 4$$
$$x + 2y \leq 12$$
$$x \leq 4$$
$$x, y \geq 0$$

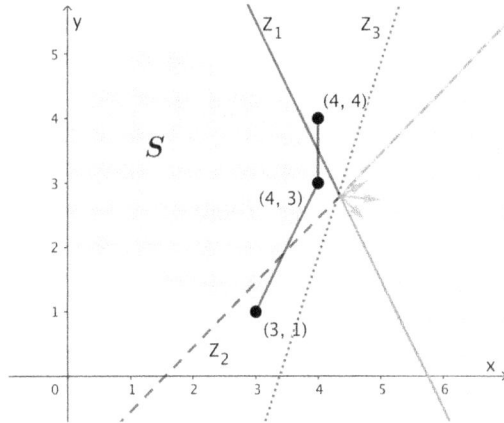

Figure 8.11: Feasible Region, Optimal Solutions, Dominance Cone and Efficient Frontier

Following the DM's preferences, the first objective function to be solved Z_3 with $(x, y) \in S$. This problem is the same as solving Z_3 independently. One already knows that the optimal solution is vertex $(4, 3)$ to which corresponds $Z_3^* = 9$. Since the DM does not want to compromise more that 20% from the optimal value of Z_3, then the second model to be solved is given below.

$$\text{Max } Z_2 = -x + y$$
$$\text{s.t. } 3x - y \geq 0.8 \cdot Z_3^*$$
$$(x, y) \in S$$

Figure 8.12 shows the reduced feasible region S_1. It also shows Z_2 optimal solution does not change (it is still vertex $(3,1)$) since the new constraint does not change the efficient frontier. Therefore, the set of interesting solutions for the DM do not suffer any changes when setting the bound of at least 80% of the optimal value[11].

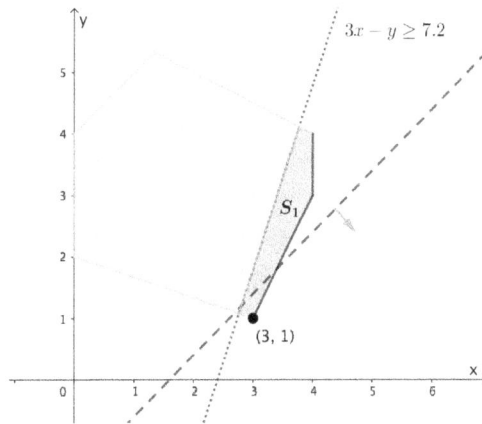

Figure 8.12: Reduced Feasible Region for the Second Objective Function

---

[11] In the traditional lexicographic approach, the analyst should continue with the method. However, this result can be shown to the DM since with this new information she might be willing to revise the value she set as being the minimum.

The next iteration will further constraint the feasible region. The model to be solved optimizes the last objective function having two constraints that will impose a limit on the values of the other two objectives. Since Z_2 is an objective to be minimized the constraint does not set a lower bound (as imposed for Z_1) but an upper bound.

The new feasible region (set S_2) is depicted in Figure 8.13. Now the efficient frontier has changed and consequently the optimal solution for Z_1 is now vertex (3.4, 1.8). Since this is the last objective function, the method has reached the compromise solution that, according, to the lexicographic approach, best reflects the DM's preferences. This solution is (3.4, 1.8) to which correspond the values of Z_1 (3.4, 1.8) = 8.6, Z_2 (3.4, 1.8) = −1.6 and Z_3 (3.4, 1.8) = 8.4.

$$\text{Max } Z_1 = 2x + y$$
$$\text{s.t.} \quad 3x - y \geq 0.8 \cdot Z_3^*$$
$$-x + y \leq 0.8 \cdot Z_2^*$$
$$(x, y) \in S$$

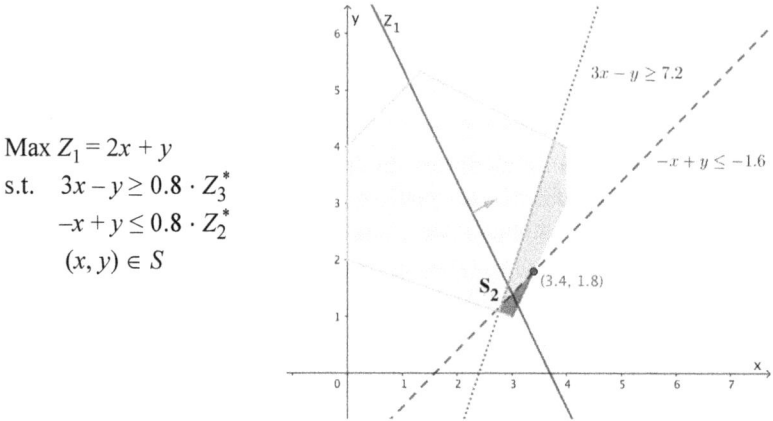

Figure 8.13: Reduced Feasible Region for the Third Objective Function.

(b) Determine the compromise solution that reflects the DM's preferences according to the ε-constraint method.

As each objective function has already been solved independently, one can proceed to formulate the model to be solved according to the ε-constraint method. Remember that one should optimize the most important objective, while all other objective functions are set as constraints. As for the lexicographic method, the new constraints should take into account if the objective is to be maximized or minimized. For the former, a lower bound is set assuring the objective to be at least some value set by the DM ($2x + y \geq 0.8 \cdot Z_1^*$). The latter comes in the form of an upper bound ($-x + y \leq 0.8 \cdot Z_2^*$). Such constraints render the problem infeasible since they do not intersect over the feasible region (Fig. 8.14). At this moment, the analyst should present this information to the DM and ask her to revise the bounds set for the constraints.

Max $Z_3 = 3x - y$
s.t. $2x + y \geq 9.6$
        $-x + y \leq -1.6$
        $(x, y) \in S$

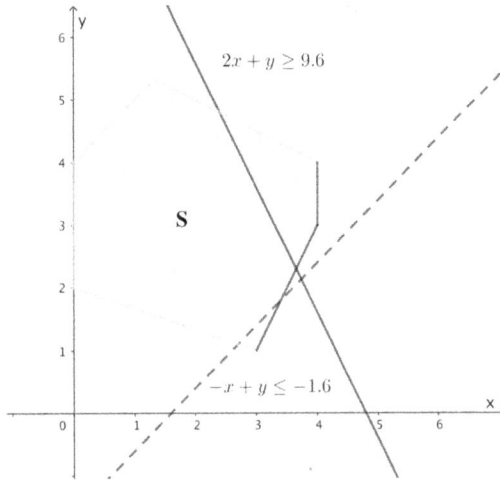

Figure 8.14: Reduced Feasible Region for the Second
Objective Function.

8.4.2   Proposed Exercises

If Chapter 7's exercises have been solved, the first question of each exercise has already been addressed.

1. Consider the MOLP problem

$$\text{Max } Z_1 = y$$
$$\text{Max } Z_2 = x - 2y$$
$$\text{s.t.} \quad 5x - 2y \leq 40$$
$$3x + 5y \leq 55$$
$$-x + y \leq 3$$
$$x + y \geq 3$$
$$x, y \geq 0$$

(a) Graph the feasible region and determine the efficient frontier.
(b) Assume the decision maker considers objective 2 to be the most important. Knowing that the decision maker, if asked, will answer that she does not wish to let go more than 25% the optimum value determine the compromise solution applying:
   (i) The lexicographic method.
   (ii) The ε-constraint method.
(c) Investigate which of the above solutions is closer to the optimal value of the most important objective.
(d) Solve exercise (b) in the objectives space.

2. Consider the MOLP problem

$$\text{Max } Z = (2x_1 + x_2; x_1 - x_2)$$
$$\text{s.t.} \qquad x_1 + x_2 \geq 2$$
$$-2x_1 + x_2 \geq 0$$
$$-x_1 + x_2 \leq 4$$
$$x_1, x_2 \geq 0$$

(a) Determine the Pareto Front.
(b) Using different bounds for the constraints, apply the lexicographic approach to solve the problem, assuming:
   (i) $Z_1 \gg Z_2$
   (ii) $Z_2 \gg Z_1$.
(c) Perform a similar analysis applying the ε-constraint method.

3. Consider the MOLP problem

$$\text{Max } Z = (2x_1 + x_2; -x_1 + x_2)$$
$$\text{s.t.} \qquad x_1 + x_2 \geq 2$$
$$2x_1 - x_2 \leq 4$$
$$-x_1 + x_2 \leq 4$$
$$x_1 + 2x_2 \leq 12$$
$$x_1, x_2 \geq 0$$

(a) Solve each objective function independently and determine the efficient frontier.
(b) Solve the problem with the lexicographic approach assuming $Z_2 \gg Z_1$.
(c) Compute the compromise solution one obtains using the ε-constraint method when Z_2 is the most important objective and the upper bound for Z_1: $Z_1 \geq 10$.

4. Consider the MOLP problem

$$\text{Max } Z_1 = -x + 3y$$
$$\text{Max } Z_2 = 3x + 3y$$
$$\text{Max } Z_3 = x + 2y$$
$$\text{s.t.} \qquad y \leq 4$$
$$x + 2y \leq 10$$
$$2x + y \leq 10$$
$$x, y \geq 0$$

(a) Graph the feasible region in the decisions space and determine which solutions might be of interest to the decision maker.
(b) Using the lexicographic method, find a compromise solution considering Z_1 as being the most important and Z_3 the least important one. Suppose

that the decision maker allows the optimal value of any objective functions to be relaxed at most by 10%.

(c) Apply the lexicographic method and find the compromise solution of each objective function's ordering. Assume the decision maker does not allow the relaxing of any objective.

   (i) $Z_1 \gg Z_2 \gg Z_3$

   (ii) $Z_1 \gg Z_3 \gg Z_2$

   (iii) $Z_2 \gg Z_3 \gg Z_1$

   (iv) $Z_2 \gg Z_1 \gg Z_3$

   (v) $Z_3 \gg Z_1 \gg Z_2$

   (vi) $Z_3 \gg Z_2 \gg Z_1$

(d) Is any of the above solutions a weakly efficient one?

(e) Knowing that the decision maker prioritizes Z_3 over all other objectives and that she allows a maximum reduction of 15% in optimal values, determine the corresponding compromise solution applying the ε-constraint method.

5. Consider the MOLP problem

$$\text{Max } Z_1 = 2x + y$$
$$\text{Max } Z_2 = 0.5x + y$$
$$\text{s.t.} \qquad x \geq 3$$
$$y \geq 4$$
$$x \leq 8$$
$$-x + y \geq 3$$
$$0.5x + y \leq 12$$

(a) Solve each objective function independently and determine the efficient frontier.

(b) Apply both lexicographic and ε-constraint methods and determine some compromise solutions, testing different ε values and different preference ordering for the objective functions.

6. Aunt Leonarda recently bought 22 kilograms of cregola, a very rare substance used in perfumery. She wants to use it to produce her two most famous perfumes, named as X and Y, for reasons of confidentiality. Aunt Leonarda wonders: "Each kiloliter of X gives me a profit of one thousand pilins, while with the perfume Y I manage to collect profits of 3 thousand pilins per kiloliter. But, each kiloliter of Y consumes two kilos of cregola, and with X the consumption is one to one."

"Don't forget that to fulfil the production plan, at least three kiloliters of X and two of Y should be produced, and that our distributor can only assure selling eight kiloliters of Y", added Blimunda, Aunt's secretary. "And there is something else, the total of X produced has to exceed the total production of Y by six kiloliters", Blimunda reminded her.

"What a mess, I want to minimize the consumption of cregola with this production, but I also want to make the highest profit." Admitted the friendly Aunt Leonarda.

Blimunda wrote something on paper and proudly presented the following formulation to Aunt Leonarda.

$$\text{Max } Z(-x_1 - x_2; x_1 + 3x_2)$$
$$\text{s.t.} \qquad x_1 \geq 3$$
$$x_2 \geq 2$$
$$x_2 \leq 8$$
$$x_1 - x_2 \leq 6$$
$$x_1 + 2x_2 \leq 21$$

Aunt Leonarda cannot decide which is the most important objective.

(a) Comment on the above formulation, establishing the relation between each constraint/objective function with Aunt's and Blimunda's observations.

(b) Using the above formulation and the ε-constraint method, determine different compromise solutions to present to Aunt Leonard. Test different ε values.

(c) Do a similar analysis using the lexicographic method.

7. Jaime is preparing his study time for next week. He will have two tests (Algebra and Biochemistry). For each study hour of Algebra his girlfriend will offer him 5 dl of strawberry ice cream. But she already threatened him that she would reduce 2 dl of ice cream for every hour he studied Biochemistry. Note that his girlfriend is a Math student. Every hour of Biochemistry study will give him 4 units of satisfaction. But an hour of Algebra study will provide him with a unit of negative satisfaction.

Jaime does not want to study more than 6 hours of Algebra, but the amount of Biochemistry study hours should not exceed in 3 hours the Algebra study period. He would like to maximize the amount of ice cream he will receive, but also the total satisfaction with the study.

This problem can be formulated as

$$\text{Max } Z_1 = 5x - 2y$$
$$\text{Max } Z_2 = -x + 4y$$
$$\text{s.t.} \qquad -x + y \leq 3$$
$$x \leq 6$$
$$x, y \geq 0$$

(a) Solve each objective function independently and determine the Pareto Front.

(b) Knowing that ice cream as the most important objective for Jaime. Determine, using the lexicographic method, the compromise solution that verifies:

(i) at least 10 dl of ice cream;

(ii) at least 24 dl of ice cream;

(iii) at least 35 dl of ice cream.

(c) Investigate if the ε-constraint method produces the same compromise solutions.

8. In a factory there are two types of machines used in the production lines of four products. The available capacities of each type of machine (in hours / week) and the number of processing hours / week required at each machine, per unit produced product, are presented in the table:

Machine	P1	P2	P3	P4	Availability
A	2	1	4	3	60
B	3	4	1	2	60

The company management wants to maximize profit, quality of P1, P2, P3 and P4 supply and the degree of employee satisfaction. Recent studies have made it possible to establish the unit contribution of each product to each objective:

	P1	P2	P3	P4
Profit	3	1	2	1
Quality	1	−1	2	4
Satisfaction	−1	5	1	2

(a) Formulate the problem as a multi-objective linear programming model.

(b) Using an optimization software (e.g. GAMS), apply the lexicographic method to determine a compromise solution for the following ordering of importance (from the most to the least important). Assume also that the decision maker allows to relax up to 10% the optimal values of the objectives.

(i) satisfaction, profit, quality,

(ii) quality, satisfaction, profit,

(iii) quality, profit, satisfaction,

(iv) profit, quality, satisfaction.

(c) Consider two objective function at a time and approximate the Pareto Front using the ε-constraint method.

(d) Apply the AUGMECON method to the three objective problems so that at least 10 Pareto non-dominated solutions are generated.

References

Antunes, C.H., M.J. Alves and J. Clímaco (2016). Multiobjective linear and integer programming. Springer, New York, NY.

Bouyssou, D., T. Marchant, M. Pirlot, A. Tsoukias and P. Vincke (2006). *Evaluation and Decision Models with Multiple Criteria: Stepping Stones for the Analyst* (Vol. 86). Springer Science & Business Media.

Ehrgott, M. and S. Ruzika (2008). Improved ε-constraint method for multiobjective programming. *Journal of Optimization Theory and Applications*, 138(3), 375.

Haimes, Y.Y., L.S. Lasdon and D.A. Wismer (1971). On a bicriterion formulation of the problems of integrated system identification and system optimization. *IEEE Trans. Syst. Man. Cybern.*, 1, 296–297.

Herzel, A., S. Ruzika and C. Thielen (2021). Approximation methods for multiobjective optimization problems: A survey. *INFORMS Journal on Computing* 33(4),1284-1299.

Laumanns, M., L. Thiele and E. Zitzler (2006). An efficient, adaptive parameter variation scheme for metaheuristics based on the epsilon-constraint method. *European Journal of Operational Research*, 169(3), 932–942.

Mavrotas, G. (2009). Effective implementation of the ε-constraint method in multi-objective mathematical programming problems. *Applied Mathematics and Computation*, 213(2), 455–465.

Mavrotas, G. and K. Florios (2013). An improved version of the augmented ε-constraint method (AUGMECON2) for finding the exact pareto set in multi-objective integer programming problems. *Applied Mathematics and Computation*, 219(18), 965–969.

Ruzika, S. and M.M. Wiecek (2005). Approximation methods in multiobjective programming. *Journal of Optimization Theory and Applications*, 126(3), 473–501.

Steuer, R.E. (1986). *Multiple Criteria Optimization: Theory, Computation and Applications*. John Wiley and Sons, Inc.

Winston, W.L. (2002). Introduction to Mathematical Programming: Applications and Algorithms, Duxbury.

Weighted Sum and Distance Minimization Methods

9.1   Introduction

The weighted sum and distance minimization methods are two approaches that fit within the group of Scalarization Methods as they reframe the Multi-Objective Linear Model (MOLP) into a single objective problem.

A weighted sum is very often the first approach that comes to mind when dealing with several perspectives in a decision-making process. This method asks the Decision Maker (DM) to assign weights to each objective, reflecting its importance to the overall problem. This assignment may be hard to do for the DM and inadequate weights might lead to a bias in the result, leading to decisions that do not reflect the DM's preferences. Other issues need to be handled with care since they may *undo* the work done to come up with the perfect weights.

The distance minimization approach seeks to find the compromise solution that is closer to some desirable reference point. This reference point may be the ideal solution or any other solution the DM wishes to attain. The question is how does one measure "close to"?

Formally, a MOLP problem can be formulated as follows:

$$\text{Max } (Z_1(x), Z_2(x), ..., Z_k(x))$$
$$\text{s.t. } x \in S$$

with x the vector of decision variables, $Z_1(x)$, $Z_2(x)$, ..., $Z_k(x)$ the k objective functions and S the feasible region defined by all model constraints[1].

Both models will be presented in detail and an example will illustrate the topic in study. Since one is seeking to find a compromise solution, Example 9.1 will be the baseline example.

---

[1] For brevity, some fundamental concepts addressed in Chapter 7 will not be defined in this chapter. However, a supplementary chapter is available for free with all the main definitions and concepts needed to fully understand the topics addressed in this chapter.

Example 9.1

Suppose the DM aims at determining the compromise solution of three objective function problems (presented below). Figure 9.1 shows the feasible region, the three objective functions and their gradient vectors.

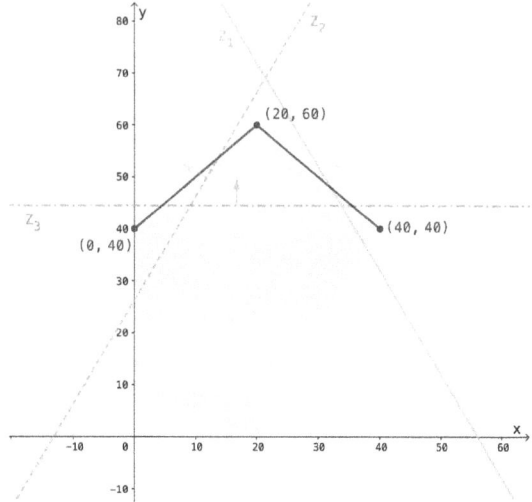

Max $Z_1 = 20x + 10y$

Max $Z_2 = -80x + 40y$

Max $Z_3 = y$

s.t. $x + y \geq 20$

$2x - y \leq 40$

$-x + y \leq 80$

$x + y \leq 80$

$x, y \geq 0$

Figure 9.1: Feasible Region and Optimal Solutions of Each Objective Function

The optimal solutions of each objective function when optimized individually[2] are

$$x_1^* = (40, 40) \text{ with } Z_1^* = 1200,$$

$$x_2^* = (0, 40) \text{ with } Z_2^* = 1600,$$

$$x_3^* = (20, 60) \text{ with } Z_1^* = 60.$$

This chapter is organized into two parts. The first one addresses the weighted sum method being the weighted sum function firstly formally defined (section 9.2.1). How to handle objective functions with different scales is addressed in section 9.2.2. Some final remarks will be drawn in the last section of the method (section 9.2.3). The second part of this chapter focuses on the distance minimization method. It starts with the definition of a distance function. Sections 9.3.1 and 9.3.2 are dedicated to the linear models that minimize the distance to a reference point, while section 9.3.3 briefly presents the quadratic model, the nonlinear model that one obtains when minimizing the Euclidean distance. Some final remarks concerning distance minimization modelling are drawn in section

---

[2] For further details on how to graphically solve a linear programming problem, the interested reader should refer to any Operations Research reference book as, for example, Winston (2002).

(section 9.3.4). This chapter ends with some overall final remarks (section 9.4) and with a set of proposed exercises (section 9.5).

9.2 Weighted Sum Model

The weighted sum model may be the most used method to deal with multiple objectives. In short, one takes all the objective functions and multiplies each of them with a weight that should reflect the importance given by the Decision Maker (DM) to each objective. The compromise solution will be the optimal solution of a single objective function that is the sum of all weighted objectives. Although being quite simple and intuitive as a method, some details need to be handled with care, otherwise the objective function may not adequately model the DM's preferences.

9.2.1 Definition of a Weighted Sum Objective Function

The weighted sum model is a single objective model where the objective function is the sum of the k original objective functions multiplied by positive values λ_p:

$$\text{Max } Z_\lambda = \sum_{p=1}^{k} \lambda_p Z_p(x)$$

$$x \in S$$

where S is the problem feasible region and λ_p define a weight vector $\Lambda = (\lambda_1, \ldots, \lambda_k)$ such as $\lambda_p > 0$, $p = 1, \ldots, k$ and $\sum_{p=1}^{k} \lambda_p = 1$.

It can be shown that the optimal solution of the above weighted sum model is an efficient solution of the MOLP problem (the interested reader might refer to Antunes et al. (2016) for more details).

Example 9.2

Suppose the DM of Example 9.1 thinks the first objective is twice more important than the other two, and these other two have equal importance. As all weights should sum one, then $\lambda_1 = \frac{1}{2}, \lambda_2 = \lambda_3 = \frac{1}{4}$. The weighted sum objective function that reflects the DM's preferences is

$$Z = \frac{1}{2}(20x + 10y) + \frac{1}{4}(-80x + 40y) + \frac{1}{4}y$$

$$\Leftrightarrow Z = -10x + \frac{61}{4}y$$

The weighted sum model is formulated below. Figure 9.2 depicts the new objective function and the compromise solution.

$$\text{Max } Z = -10x + \frac{61}{4}y$$

$$\text{s.t.} \quad x + y \geq 20$$

$$2x - y \leq 40$$

$$-x + y \leq 40$$

$$x + y \leq 80$$

$$x, y \geq 0$$

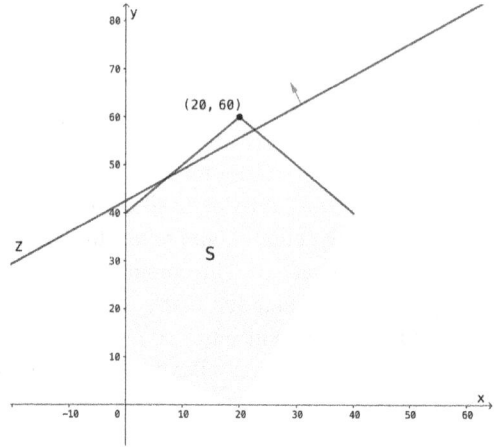

Figure 9.2: Feasible Region and Optimal Solution

of the Weighted Sum with $\Lambda = \left(\dfrac{1}{2}, \dfrac{1}{4}, \dfrac{1}{4}\right)$

The optimal solution is vertex $(20, 60)$[3] and the corresponding objective function values are $Z_1(20, 60) = 1000$, $Z_2(20, 60) = 1600$ and $Z_3(20, 60) = 60$.

However, the optimal solution of the weighted function (the compromise solution) can very much be influenced by the scales of the objective functions. In the previous example functions Z_1 and Z_2 are 10-times "heavier" than the third objective function. Could this have influenced the optimal solution? The next section will be dedicated to methods on how to deal with objective functions of different magnitude.

9.2.2   Dealing with Functions in Different Scales

When objective functions have different scales, prior to the application of the weighted sum method, one should rescale all functions so that they have the same magnitude (Steuer 1986). Several different normalization methods have been proposed in the literature. In this chapter, three different approaches will be addressed. However, this is not an exhaustive list. For instance, Miettinen (2008) suggests using the objective function range over the Pareto Front, calculated with the nadir and ideal solutions.

To better understand the impact of objective functions with different scales have on the compromise solution proposed by the method, the Example 9.3 model will be first solved without taking the objectives' function magnitude into consideration.

---

[3] Notice that this compromise solution was never proposed by any of the methods addressed in Chapter 8. This reinforces the idea that each method may produce its compromise solutions.

Example 9.3

Consider a three-objective MOLP model defined in S. Fig. 9.3 depicts all three objective functions and the corresponding optimal solutions when solved independently.

Max $Z_4 = 24x + 51y$

Max $Z_5 = 4x + 3y$

Max $Z_6 = -4700x + 3250y$

s.t. $x + y \geq 20$

$2x - y \leq 40$

$-x + y \leq 40$

$x + y \leq 80$

$x, y \geq 0$

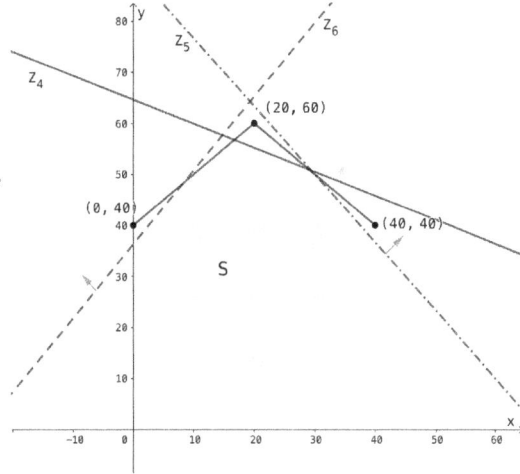

Figure 9.3: Feasible Region, Optimal Solutions of each Objective Function, and Pareto Front.

When solved independently, vertex $x_4^* = (20, 60)$ is the optimal solution of Z_4, which corresponds to the optimal value of $Z_4^* = 3540$. The optimal solution of Z_5 is $x_5^* = (40, 40)$ with $Z_5^* = 280$, and the optimal solution of Z_6 is $x_6^* = (0, 40)$ with $Z_6^* = 130000$.

The DM considers that Z_5 is the most important objective, followed by Z_4 and with the least importance is Z_6, i.e. $Z_5 \gg Z_4 \gg Z_6$, where "$\gg$" reads as "more important than". The weight vector $\Lambda = (0.4, 0.5, 0.1)$ has been agreed by the DM as reflecting her preferences. Ignoring the different scales, the weighted sum function is

$$\text{Max } Z = -458.4x + 346.9y .$$

With such an objective function, the compromise solution is vertex $(0, 40)$ and the values for each of the objective function are $Z_4 (0, 40) = 2040$, $Z_5 (0, 40) = 120$ and $Z_3 (0, 40) = 130\ 000$.

It is not a coincidence that the previous compromise solution is also the optimal solution of objective Z_6. Although this is the least important objective function with a weight five times smaller than the most important objective, the difference in scale among objectives makes Z_6 as the one with the highest "power of attraction". Therefore, it "pulls" the weighted sum function towards its optimal solution. So, what should be done before building the weighted sum functions? The most common approach is to scale all objective functions (commonly called as *normalization*[4] although this is one of the scaling methods). Three different approaches will be addressed: normalization, an adequate power of 10, and the range equalization factor.

9.2.2.1 Normalization

Normalization, as the name suggests, implies the use of *norms* to scale each function. A *norm* is a function that measures the length of vectors[5] . Formally, a **norm** is any function[6] $|| \cdot ||$: $R^n \to R$ defined over vector space R^n, where for any vectors $x, y \in R^n$ and scalar $\alpha \in R$, satisfies the following properties:

- $||x|| \geq 0$
- $||x|| = 0 \Leftrightarrow x = 0$
- $||\alpha x|| = |\alpha| \, ||x||$
- $||x + y|| \leq ||x|| + ||y||$

Among the norms there are the L_p-norms. This is a family of norms that are defined as

$$|| \, x \, ||_p = \sqrt[p]{\sum_{i=1}^{n} |x_i|^p}$$

with $p \in \{1, 2, ...\} \cup \{\infty\}$ and $x = (x_1, x_2, ..., x_n)$. When $p = 1, 2$ and ∞ one has

$$|| \, x \, ||_1 = \sum_{i=1}^{n} |x_i|$$

$$|| \, x \, ||_2 = \sqrt{\sum_{i=1}^{n} |x_i|^2}$$

$$||x||_\infty = \max (|x_1|, |x_2|, ..., |x_n|)$$

These norms are so common that they have been named. $|| \cdot ||_1$ is known as the Manhattan norm since it is related with the distance between two points in a rectangular grid. $|| \, x \, ||_2$ is the Euclidean norm and $|| \, x \, ||_\infty$ is the Tchebychev norm and its expression comes from the fact that the largest component (in absolute value) dominates over all others when raised to the power of $p = \infty$. Norms are used to define the distance between two points (section 9.3.1).

---

[4] In Chapter 3, when studying the SMART methodology (also a weighted sum approach), the scaling/normalization step has been shown as essential to perform an adequate modelling.

[5] It will be assumed that all vectors are defined in R^n.

[6] Norms are functions defined in any vector space. However, in this book, only R^n is of interest.

Example 9.4

Consider the vector $x = (2, -3, 1)$. The L_1, L_2 and L_∞-norms of x are

$$\|x\|_1 = |2| + |-3| + |1| = 6$$

$$\|x\|_2 = \sqrt{|2|^2 + |-3|^2 + |1|^2} = \sqrt{14} = 3.74$$

$$\|x\|_\infty = (|2|, |-3|, |1|) = 3$$

Dividing each component of the vector by its norm will compute a new vector of length 1 (of norm equal to 1) which is the *normalized* vector. So, normalizing vector x according to each norm one gets

$$\frac{x}{\|x\|_1} = \frac{1}{6}(2, 3, 1) = \left(\frac{1}{3}, \frac{1}{2}, \frac{1}{6}\right)$$

$$\frac{x}{\|x\|_2} = \frac{1}{3.74}(2, 3, 1) = (0.53, 0.8, 0.27)$$

$$\frac{x}{\|x\|_\infty} = \frac{1}{3}(2, 3, 1) = \left(\frac{2}{3}, 1, \frac{1}{3}\right)$$

Which is then the weighted sum model of Example 9.3, when the objective functions are scaled using this normalization procedure? The three examples below apply each of the norms to the MOLP model to determine the weighted sum compromise solution considering the weights $\Lambda = (0.4, 0.5, 0.1)$. Note that, as defined in Chapter 7, each function can be viewed as the product of two vectors: the coefficient vector and the variables' vector. For instance, $Z = 2x_1 - 3x_2 + x_3$ is written in the vector form as $Z = c^T x$, with $c^T = (2, -3, 1)$ and $x^T = (x_1, x_2, x_3)$. The normalization procedure is applied to vector c.

Example 9.5

Taking the Manhattan norm (L_1-norm), the first step is to compute the norm of each objective function:

$$\|Z_4\| = |24| + |51| = 75$$

$$\|Z_5\| = |4| + |3| = 7$$

$$\|Z_6\| = |-4700| + |3250| = 7950.$$

The normalized functions are then

$$Z_4' = 0.32x + 0.68y$$

$$Z_5' = 0.57x + 0.43y$$

$$Z_6' = -0.59x + 0.41y$$

One quick check one can do at this moment is to see if the sum of the objective functions' coefficients, in absolute value, equals 1. If not, there is a miscalculation somewhere. The weighted sum function is then

$$Z = 0.4 \cdot Z_4' + 0.5 \cdot Z_5' + 0.1 \cdot Z_6'$$
$$= 0.354x + 0.528y$$

Figure 9.4 shows the compromise solution, vertex $x^* = (20, 60)$, which corresponds with $Z_4 (20, 60) = 3540$, $Z_5 (20, 60) = 260$ and $Z_6 (20, 60) = 101000$. Now the optimal solution is no longer the optimal solution of the least important function. It is now "closer" to the optimal solution of the most important objective.

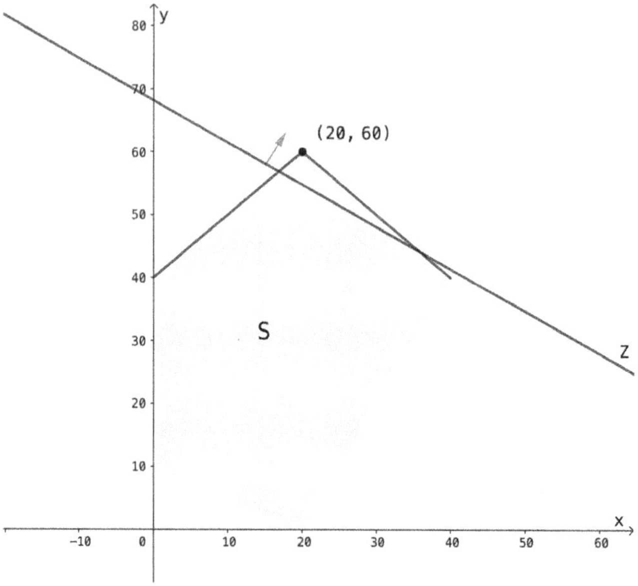

Figure 9.4: Objective Function and Optimal Solution for the
Weighted Sum Function Normalized with L_1-norm.

Example 9.6

Taking the Euclidean norm (L_2-norm), the first step is to compute the norm of each objective function:

$$\| Z_4 \| = \sqrt{|24|^2 + |51|^2} = 56.36$$

$$\| Z_5 \| = \sqrt{|4|^2 + |3|^2} = 5$$

$$\| Z_6 \| = \sqrt{|4700|^2 + |3250|^2} = 5714.24$$

The normalized functions are then

$$Z_4' = 0.43x + 0.9y$$
$$Z_5' = 0.8x + 0.6y$$
$$Z_6' = -0.82x + 0.57y$$

The weighted sum function is then

$$Z = 0.4 \cdot Z_4' + 0.5 \cdot Z_5' + 0.1 \cdot Z_6'$$
$$= 0.49x + 0.717y$$

The compromise solution is the same as for the L_1-norm (Figure 9.5). However, this may not always be the case. Different norms may produce different compromise solutions. In this example, both normalized weighted sum objective functions have very similar gradient vectors.

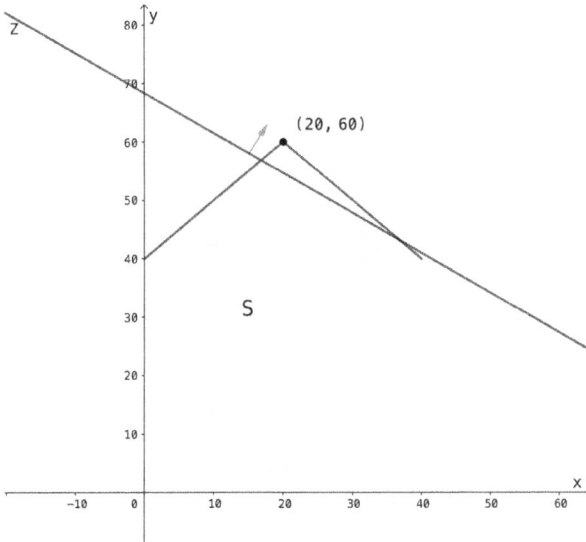

Figure 9.5: Objective Function and Optimal Solution for the Weighted Sum Function Normalized with L_2-norm.

Example **9.7**

Taking the Tchebychev norm (L_∞-norm), the first step is again to compute the norm of each objective function:

$$\|Z_4\| = \max(|24|, |51|) = 51$$
$$\|Z_5\| = \max(|4|, |3|) = 4$$
$$\|Z_6\| = \max(|-4700|, |3250|) = 4700.$$

The normalized functions are then

$$Z_4' = 0.47x + y$$
$$Z_4' = x + 0.75y$$
$$Z_6' = -x + 0.69y$$

The weighted sum function is then

$$Z = 0.4 \cdot Z_4' + 0.5 \cdot Z_5' + 0.1 \cdot Z_6'$$
$$= 0.588x + 0.844y$$

The compromise solution is again vertex (20, 60), since the gradient vector has only changed slightly (Fig. 9.6). As mentioned in the previous example, this is not always the case.

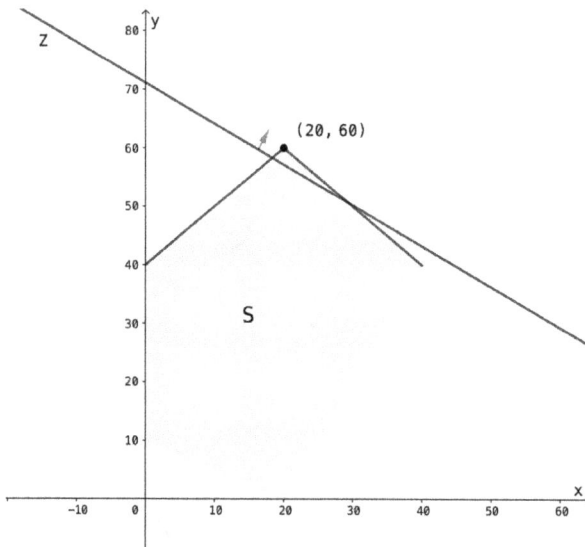

Figure 9.6: Objective Function and Optimal Solution for the Weighted Sum Function Normalized with L_∞-norm.

In larger dimensional problems, different metrics may easily lead to different compromise solutions, as the solution space is usually a polyhedron with a larger number of facets and consequently of vertices. Example 9.8 shows a model where the compromise solutions are not only different for when using different norms, but they also differ from the optimal solutions of each objective when solved independently.

Example 9.8

Consider the MOLP model below.

$$\text{Max } Z_1 = 74500x_1 + 299700x_2 - 64500x_3 + 19300x_4$$
$$\text{Max } Z_2 = 10.16x_1 - 3.25x_2 + 8.67x_3 + 4.94x_4$$
$$\text{Max } Z_3 = -2036x_1 - 3290x_2 + 1230x_3 + 7300x_4$$
$$\text{s.t.} \quad x_1 + x_2 + x_3 + x_4 \leq 250$$
$$-x_1 + x_2 - x_3 \leq 190$$
$$-x_1 + x_2 - x_4 \leq 150$$
$$x_1 + x_4 \leq 180$$
$$x_1, x_2, x_3, x_4 \geq 0$$

The optimal solutions when solving each objective function independently and the corresponding values in all objectives are given in the payoff table (Table 9.1).

Table 9.1: Payoff Table

	Z_1	Z_2	Z_3
$x_1^* = (50,200,0,0)$	$63665*10^3$	-142	$-759.8*10^3$
$x_2^* = (180,0,70,0)$	$8895*10^3$	2435	$-280.38*10^3$
$x_3^* = (0,0,70,180)$	$-1041*10^3$	1496.1	$1400.1*10^3$

When taking each of the three norms to rescale the objective functions, the weighted sum model produces two different compromise solutions (Table 9.2) none of them is an optimal solution presented in the payoff table.

Table 9.2: Compromise Solutions for the Weighted Sum Model Scaled with Different Norms and Corresponding Objective Functions Values

Norm	Compromise solution	Z_1	Z_2	Z_3
L_1 and L_2	$(0,190,0,60)$	$58\ 101*10^3$	-321.1	$-187.1*10^3$
L_∞	$(0,70,0,180)$	$24\ 453*10^3$	661.7	$1083.7*10^3$

Note that the objective function values presented in Table 9.2 are the non-scaled values as these are the one that make sense to the DM.

To the question "what is the best norm to use?", there is no definite answer. The use of each one of them is formally correct and, therefore, the best one is the one that produces the compromise solution that best reflects the DM's preferences.

One of the disadvantages of this scaling procedure is that it produces changes on the coefficients which makes them unrecognizable numbers to the DM. This drawback is overcome by the next procedure.

9.2.2.2. An Adequate Power of 10

As the name suggests, this procedure scales all objectives' to the same unit. Firstly, the unit of the weight sum objective function needs to be chosen. Then, for each of the functions, one determines on the power of 10 that adequality renders the function to the chosen unit. Therefore, the weighted sum function will be the sum of functions, all in the same unit.

Example 9.9

Taking the MOLP model of Example 9.3 one can see the unit of Z_4 is 10 times the unit of Z_5, while Z_6 is 10^4 the unit of Z_5. Taking Z_5 as the base unit one should rescale Z_4 and Z_6:

$$Z_4' = \frac{Z_4}{10} = 2.4x + 5.1y$$

$$Z_6' = \frac{Z_6}{10^4} = 4.7x + 3.25y$$

The two new functions, Z_4' and Z_6', will be added to Z_5 defining the weighted sum function:

$$Z = 0.4 \cdot Z_4' + 0.5 \cdot Z_5 + 0.1 \cdot Z_6'$$
$$= 2.49x + 3.865y$$

Remember that the weighted vector is $\Lambda = (0.4, 0.5, 0.1)$.

Figure 9.7 shows the weighted sum objective function and the corresponding compromise solution (vertex (20, 60)).

This scaling method brings all objective functions into the same order of magnitude and maintains the recognizability of the objective function coefficients since it "only moves" the decimal point.

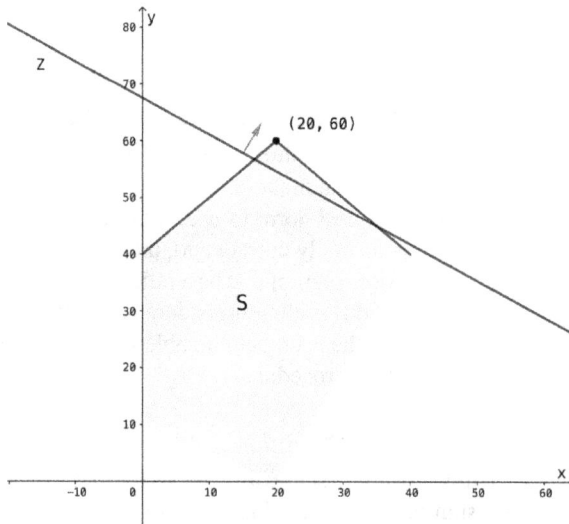

Figure 9.7: Objective Function and Optimal Solution for the Weighted Sum Function.

9.2.1.1 The Range Equalization Factor

The last procedure takes into account the variability of the objective function over the feasible region. This is done by multiplying each function by an equalization factor given by

$$\pi_i = \frac{1}{R_i}\left[\sum_{p=1}^{k}\frac{1}{R_p}\right]^{-1}$$

where R_i is the difference between the maximum and minimum values of objective function i over the feasible region.

***Example* 9.10**

Taking the MOLP model of Example 9.3, one should start by determining the minimum value over set S for each of the three objective functions[7]. The maximum values have already been calculated. They are the optimal values of each objective function when solved independently. In S, one has $Z_4 \in [510, 3540]$, $Z_5 \in [30, 280]$ and $Z_6 \in [-94\,000, 130\,000]$. Thus,

$$R_4 = 3030,\ R_5 = 250 \text{ and } R_6 = 224\,000.$$

Hence,

$$\pi_4 = 0.0762,\ \pi_5 = 0.9228 \text{ and } \pi_6 = 0.001.$$

The new objective functions are

$$Z_4' = \pi_4 \cdot Z_4 = 1.827x + 3.883y$$
$$Z_5' = \pi_5 \cdot Z_5 = 3.691x + 2.768y$$
$$Z_6' = \pi_6 \cdot Z_6 = -4.841x + 3.347y.$$

The compromise solution will be found by solving the weighted sum function in the feasible region S:

$$Z = 0.4 \cdot Z_4' + 0.5 \cdot Z_5' + 0.1 \cdot Z_6'$$
$$= 2.0922x + 3.2719y.$$

Figure 9.8 shows the weighted sum objective function and the corresponding compromise solution (again vertex (20, 60)).

For this example, no matter the scaling method applied, the compromise solution is always the same. However, this is due to the problem being a toy problem and only having three extreme points that are efficient solutions. As shown in Example 9.8, in larger problems with Pareto Fronts with a higher number of facets, the compromise solution will most certainly be different for

---

[7] These values differ from the one presented in the payoff table since the latter is an approximation of the minimum value over the Pareto Front. For more details refer to Chapter 7.

different methods. So, which should be the solution proposed to the DM? All of them can be proposed and the DM should choose the one that (s)he believes better reflects his/her preferences.

9.2.3 *Final Remarks*

This section presents the weighted sum model (seminal work of Gaas and Satty, 1955). This is a parametric modelling approach which becomes a linear optimization model for any fixed value of the parameter (the weights). When all weights are positive and the optimal solution is unique, one is assured to have reached an efficient solution. Weakly efficient solutions may appear if any weight is set to zero or the weighted sum function has multiple optimal solutions (Cohon, 2003).

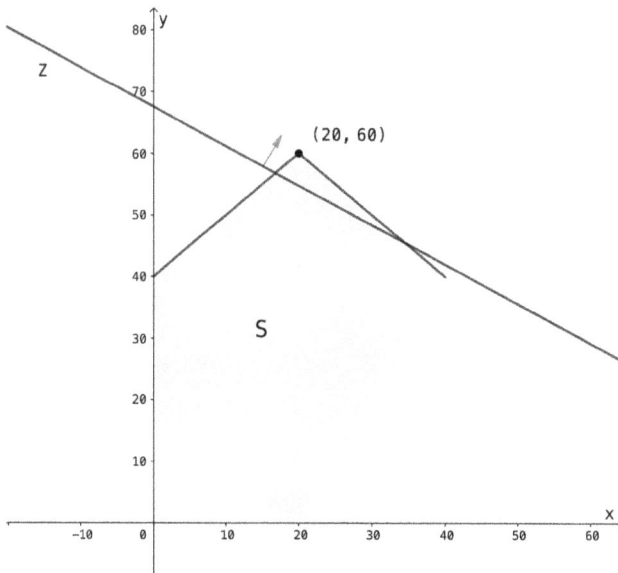

Figure 9.8: Objective Function and Optimal Solution for the Weighted Sum Function.

Before determining the weighted sum function method, a scaling procedure should be applied when the coefficients are not defined in the same order of magnitude among objective functions. If such a procedure is not applied, the methods may be fooled by "heavier" scale objectives, pushing the weighted sum optimal solution towards their optimal vertices. Three different approaches have been proposed. None of them is better than the others, and none of them is the worst. Each of them has weaknesses that are overcome by the others. For instance, if the feasible region *S* is an unbound set, the minimum value over *S* may not exist. Consequently, the range equalization factor method cannot be applied. Moreover, the choice of the method affects the compromise solution obtained.

This weighted sum method can be used to approximate the Pareto Front. To do so, several weights are used to allow determining different efficient solutions. In this case, the method can be viewed as an "a posteriori" method, since after reaching different compromise solutions, the DM may choose the one the best fits his/her preferences. However, one should be aware that the weighted sum method only produces extreme solutions (vertices of the feasible region) as compromise solutions. It is not able to determine other compromise solutions (as any solution between two adjacent vertices[8]). Cohon (2003) presents a strategy to approximate the Pareto Front by systematically varying the weights. With such a procedure, for each efficient solution generated one must, at least, solve one linear model. Quite often several sets of weights will produce the same solution. However, the use of the weighted sum method as a Pareto Front approximation method presents some limitations when compared to the ε-constraint method addressed in Chapter 8. For instance, the weighted sum method is only able to produce extreme point solutions and, consequently, a large number of weights may lead to the same solution while the ε-constraint method is able to determine a non-extreme solution and, by properly adjusting the ε values, one is able to control the number of generated solutions (Mavrotas 2009).

9.3 Distance Minimization Model

The distance minimization model aims at finding the compromise solution, among the non-dominated solutions, that is nearer to a reference solution. This latter solution might be provided by the DM, since very often (s)he have an idea of what is a "good" solution. It can also be the ideal solution, defined in Chapter 7. As the ideal solution is not a feasible solution[9] , a good compromise solution could be one that is the "closest" one. But why is it closest placed between quotation marks? Terms such as "closer" and "nearer" imply the concept of distance. In fact, when aiming to find a solution that is the "closest to" or the "nearest to", one needs to find the solution that minimizes the distance to the reference point.

Let's start by formally defining what a distance is. **Distance** (or *metric*) is defined as a function that assigns a scalar to each pair of point x, y in the vector space R^n, $d(x, y)$: $R^n \rightarrow R_0^+$ such that the following properties are satisfied:

- $d(x, y) = 0$ *if and only if* $x = y$,
- $d(x, y) = d(y, x)$,
- $d(x, y) \leq d(x, z) + d(z, y)$.

Norm functions induce distance functions. So, L_p norms (addressed in section 9.2.2) induce L_p distance functions given by

---

[8] Adjacent vertices are vertices connected by one single edge (e.g. the extreme points of a line segment).

[9] Otherwise the multi-objective problem would not exist since at least one solution would be optimal for all objectives.

$$\| x - y \|_p = \sqrt[p]{\sum_{i=1}^{n} | x_i - y_i |^P}, \quad p = 1, 2 \ldots$$

$$\|x - y\|_\infty = \max (|x_1 - y_1|, |x_2 - y_2|, \ldots, |x_n - y_n|)$$

As for the norms, three of these L_p distance functions are especially important. They are, respectively, the Manhattan distance, the Euclidean distance and the Tchebychev distance.

$$\| x - y \|_1 = \sum_{i=1}^{n} | x_i - y_i |$$

$$\| x - y \|_2 = \sqrt{\sum_{i=1}^{n} | x_i - y_i |^2}$$

$$\|x - y\|_\infty = |x_i - y_i|$$

Example 9.11

Determine the distance between $x = (1, -2, 4)$ and $y = (2, 1, 3)$, according to the three distance functions defined above:

$$\|x - y\|_1 = |1 - 2| + |-2 - 1| + |4 - 3| = 5$$

$$\| x - y \|_2 = \sqrt{|1 - 2|^2 + |-2 - 1|^2 + |4 - 3|^2} = \sqrt{1 + 9 + 1} = \sqrt{11}$$

$$\|x - y\|_\infty = (|1 - 2|, |-2 - 1|, |4 - 3|) = 3.$$

The **contour line** of a function is the line (not necessarily a straight line) connecting the points for which function has the same value. Do distances induced by different norms have different contour lines? Figure 9.9 shows the contour lines for the distances L_1, L_2 and L_∞ connecting all points whose distance function value is equal to one. So, one can easily conclude that different distance functions almost always render different points when computing the same distance value (as seen in Example 9.11).

To find the compromise solution (a solution on the Pareto Front) that is closer to the reference point (which will be assumed from now on that is the ideal solution) one needs to choose the distance function to use. The next three subsections will address each the three L_p distances alongside with an example to show how to formulate the distance model and the geometry behind them.

9.3.1   Distance Minimization to the Ideal Solution

Formally, the model that minimizes the distance to the ideal solution $Z^* = (Z_1^*, Z_2^*, \ldots, Z_k^*)$ taking the L_1-distance function can be formulated as

$$\text{Min } |Z_1(x) - Z_1^*| + |Z_2(x) - Z_2^*| + \ldots + |Z_k(x) - Z_k^*|$$
$$\text{s.t.} \qquad x \in S$$

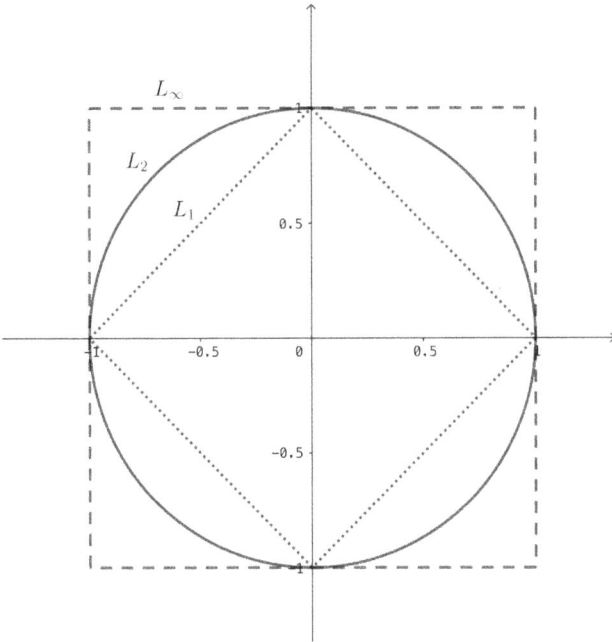

Figure 9.9: Contour Lines for Distances L_1, L_2 and L_∞ Connecting
Points Distancing One from the Origin.

with x the vector of decision variables, $Z_1(x)$, $Z_2(x)$, ..., $Z_k(x)$ the k objective
functions and S the feasible region defined by all model constraints.

Although non-linear, the L_1-distance function allows the modelling of a linear
objective function since one knows that Z_p^*, $p = 1$, ..., k is the highest value one
can obtain for $x \in S$. Then, $Z_p(x) \le Z_p^*$, $\forall\, x \in S$ which allows the "removing" of
the absolute value function. The linear model that is equivalent to the problem
above is

$$\text{Min } Z_1^* - Z_1(x) + Z_2^* - Z_2(x) + \ldots + Z_k^* - Z_k(x)$$

s.t. $x \in S$

Let's determine the non-dominated solution that minimizes the distance to
the ideal solution according to the L_1-distance function for the problem given in
Example 9.1.

Example 9.12

Consider the three objective functions

$$\text{Max } Z_1 = 20x + 10y$$
$$\text{Max } Z_2 = -80x + 40y$$
$$\text{Max } Z_3 = y$$

Given that the ideal solution is $Z^* = (1200, 1600, 60)$, the objective function that minimizes the L_1-distance to this reference point is given by

$$Z = |Z_1\,(x) - Z_1^*| + |Z_2\,(x) - Z_2^*| + |Z_3\,(x) - Z_3^*|$$
$$= |20x + 10y - 1200| + |-80x + 40y - 1600| + |y - 60|$$
$$= 1200 - (20x + 10y) + 1600 - (-80x + 40y) + 60 - y$$
$$= 2860 + 60x - 49y.$$

The model is then

Min $Z = 2860 + 60x - 49y$
s.t. $x + y \geq 20$
          $2x - y \leq 40$
          $-x + y \leq 40$
          $x + y \leq 80$
          $x, y \geq 0$

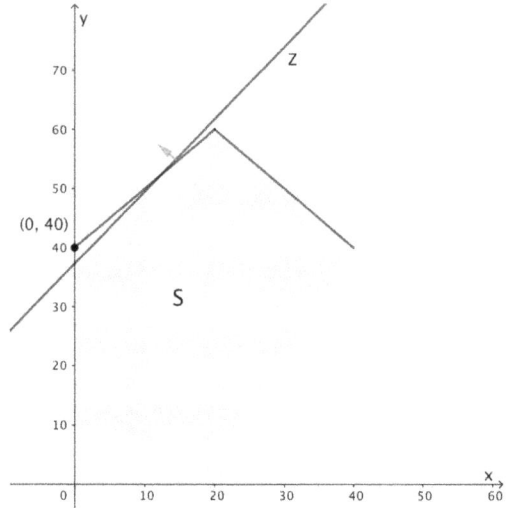

Figure 9.10: Compromise Solution When Minimizing the L_1-Distance to the Ideal Solution.

The compromise solution is vertex $(0, 40)$ which corresponds to optimal solution of Z_2 (Figure 9.10). In larger dimension problems, the compromise solution does not have to be one of the objective function optimal solutions. In this small example it will always be one of the three optimal solutions since the Pareto Front only has 3 vertices and each of them is the optimal solution of one of the objectives. One thing is certain, the compromise solution will always be a vertex of the feasible region.

Before finishing the example, remember that, since Max $F = -$ Min $- F$ and Max $F + k = k +$ Max F, where k is a constant, then

Min $2860 + 60x - 49y \Leftrightarrow 2860 -$ Max $(-60x + 49y)$.

An interesting perspective to understand the distance minimization method is to look at its geometry in the objectives' space. The geometry of the contour line allows for an intuitive knowledge of where the efficient solution lies in the Pareto Front. However, this perspective is limited to two and, with some ingenuity, to three objective functions MOLP problems.

***Example* 9.13**

Consider two of the three objective functions, Z_1 and Z_2 and feasible region S defined in Example 9.1

$$\text{Max } Z_1 = 20x + 10y$$
$$\text{Max } Z_2 = -80x + 40y$$

Figure 19.11 shows the feasible region on the objectives space (S_Z is the projection of polyhedron S onto the objectives space[10]), the ideal solution (marked with a cross; notice that it is not a feasible solution since it lies outside the feasible region) and the contour line formed by the equidistance points with respect to L_1-distance. In this case, the non-dominated solution is vertex (400,

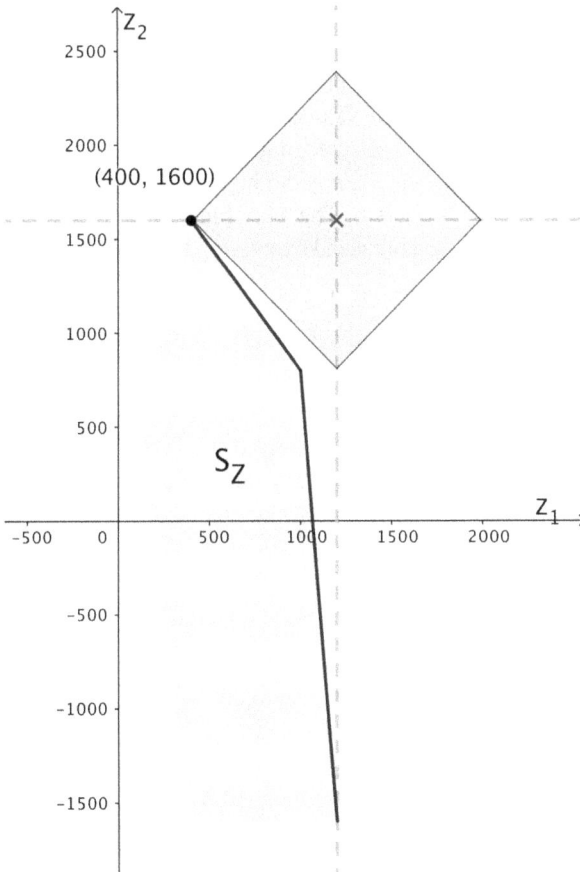

Figure 9.11: Feasible Region on the Objectives Space, Ideal Solution and Compromise Solution that Minimizes the L_1-Distance to the Ideal Solution.

---

[10] For further detail refer to Chapter 7.

1600) which is the first non-dominated point reached by the contour line when it increases the distance from the ideal solution. In the solution space it corresponds to the efficient solution $x^* = 0$ and $y^* = 40$. This graphical solution approach of the problem needs always to be validated by solving the corresponding model.

9.3.2 L_∞-Distance Minimization to the Ideal Solution

Formally, the model that minimizes the distance to the ideal solution $Z^* = (Z_1^*, Z_2^*, ..., Z_k^*)$ taking the L_∞-distance function can be formulated as

$$\text{Min Max } (|Z_1(x) - Z_1^*|, |Z_2(x) - Z_2^*|, ..., |Z_k(x) - Z_k^*|)$$
$$\text{s.t. } x \in S$$

with x the vector of decision variables, $Z_1(x)$, $Z_2(x)$, ..., $Z_k(x)$ the k objective functions and S the feasible region defined by all model constraints.

The L_∞-distance function is not a linear function. In fact, it presents two nonlinear factors: (1) the absolute value and (2) the Max function. However, both can be linearized. As in the L_1-distance model, the absolute value function can be "removed" since Z_p^*, $p = 1, ..., k$ is the highest value one can obtain for $x \in S$, therefore, $Z_p(x) \leq Z_p^*$, $\forall x \in S$. The Max function can also be linearized by introducing an auxiliary variable v and expressing the model as

$$\text{Min } v$$
$$\text{s.t. } Z_p^* - Z_p(x) \leq v, \quad p = 1, ..., k$$
$$x \in S$$

These new k constraints assure that variable v is greater or equal to $Z_p^* - Z_p$ (x), for all $p = 1, ..., k$. Minimizing v will drive down the maximum of these expressions. Does one need to impose any domain constraint to variable v? The answer is "no" since, in this case, one is sure that $v \geq 0$.

Let's determine the non-dominated solution that minimizes the distance to the ideal solution according to the L_∞-distance function for the problem given in Example 9.1.

***Example* 9.14**

The model that minimizes the L_∞-distance to the ideal solution, $Z^* = (1200, 1600, 60)$, is formulated as

$$\text{Min } v$$
$$\text{s.t.} \quad 1200 - (20x + 10y) \leq v$$
$$1600 - (-80x + 40y) \leq v$$
$$60 - y \leq v$$
$$x \in S$$

This model cannot be solved graphically since it has 3 variables. This time one has to resort to optimization software to solve it. The compromise solution is $(x^*, y^*) = (11.43, 51.43)$ which corresponds to the non-dominated solution $(742.9, 1143, 51.43)$.

In this example, the compromise solution is an efficient one but different from the ones proposed by all previous methods. This shows that different distances may propose different compromise solutions but always efficient ones.

Example 9.15 below depicts the contour line "working" in the objectives space in search for the non-dominated solution in the Pareto Front.

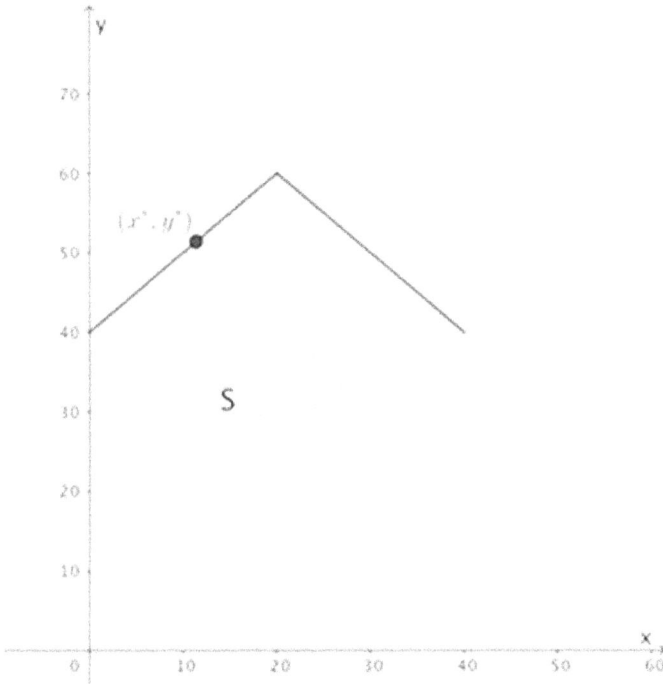

Figure 9.12: Compromise Solution (X) When Minimizing the L_1-Distance to the Ideal Solution.

Example 9.15

Again, consider two of the three objective functions, Z_1 and Z_2 and feasible region S defined in Example 9.1

$$\text{Max } Z_1 = 20x + 10y$$
$$\text{Max } Z_2 = -80x + 40y$$

Figure 9.13 shows the feasible region on the objective space, the ideal solution (marked with a cross) and the contour line formed by the equidistant points given by L_∞-distance. Now, the contour line forms a square. When "growing" to be closer to the Pareto Front it will "touch" the feasible region at the non-dominated solution (742.9, 1143). Again, be aware that this geometric capability is not an adequate tool to determine the compromise solution. One should always resort to algorithms to compute the model optimal solution.

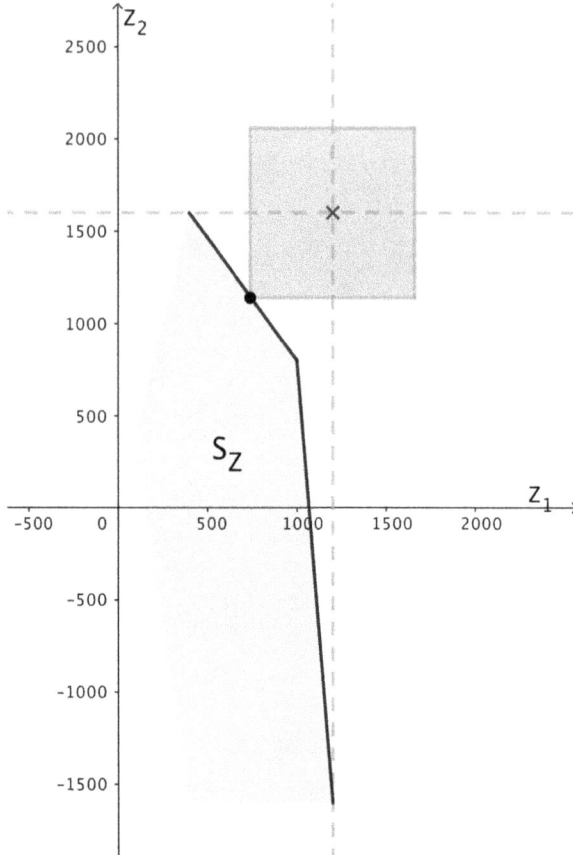

Figure 9.13: Feasible Region on the Criterion Space, Ideal Solution and Compromise Solution that Minimizes the L_∞-Distance to the Ideal Solution.

9.3.3 L_2-Distance Minimization to the Ideal Solution

Formally, the model that minimizes the distance to the ideal solution $Z^* = (Z_1^*, Z_2^*, ..., Z_k^*)$ taking the L_2-distance function can be formulated as

$$\text{Min } \sqrt{|Z_1(x) - Z_1^*|^2 + |Z_2(x) - Z_2^*|^2 + ... + |Z_k(x) - Z_k^*|^2}$$

$$\text{s.t. } x \in S$$

with x the vector of decision variables, $Z_1(x)$, $Z_2(x)$, ..., $Z_k(x)$ the k objective functions and S the feasible region defined by all model constraints.

Again L_2-distance function is not a linear function, and this time it cannot be linearized as the two previous ones. Notice, however, that

$$Min\sqrt{F(x)} = \sqrt{Min\ F(x)}$$

when $x \in R_+^n$ and, therefore, this model can be solved as a quadratic objective function as both will have the same optimal solution. The properties of quadratic functions in terms of convexity make the quadratic model "quite easy" to be solved by many optimization softwares designed to address nonlinear models. Although non-linear optimization is outside the scope of this book, a brief presentation of this model is provided since this is the model that formulates the most common distance, the Euclidean distance.

Example 9.16

The model that minimizes the L_2-distance to the ideal solution, $Z^* = (1200, 1600, 60)$, is formulated as

$$Min\ (20x + 10y - 1200)^2 + (-80x + 40y - 1600)^2 + (y - 60)^2$$
$$\text{s.t.} \quad x \in S$$

Geometrically, the contour line forms a circumference and the compromise solution is the first non-dominated solution "touched" by the circumference (Figure 9.14).

9.3.4 Final Remarks

This section presents the distance minimization model where the compromise solution is obtained by minimizing the distance to a reference point. A natural choice for the reference point is the ideal solution. This solution allows the linearization of the objective functions formulated with the Manhattan (L_1-distance) or the Tchebychev distances (L_∞-distance). If any other reference solution is chosen, close attention should be taken concerning the linearization step, as the argument that the reference point's value is larger than any feasible solution may not hold.

The choice of the distance function may change the compromise solution (Miettinen, 1998). Although it is not addressed in this section, the magnitude of the objective functions may also cause distortions in the determination of the compromise solution. Therefore, rescaling strategies, as the ones proposed when addressing the weighted sum model, should be applied.

With respect to weakly efficient solutions, L_1 and L_2 distance models produce only efficient solutions, while the Tchebychev distance (L_∞-distance) model may converge to a weakly efficient solution (Miettinen, 2008). This limitation is overcome if the (traditional) Tchebychev distance is replaced by its augmented weighted version. The model formulation is then

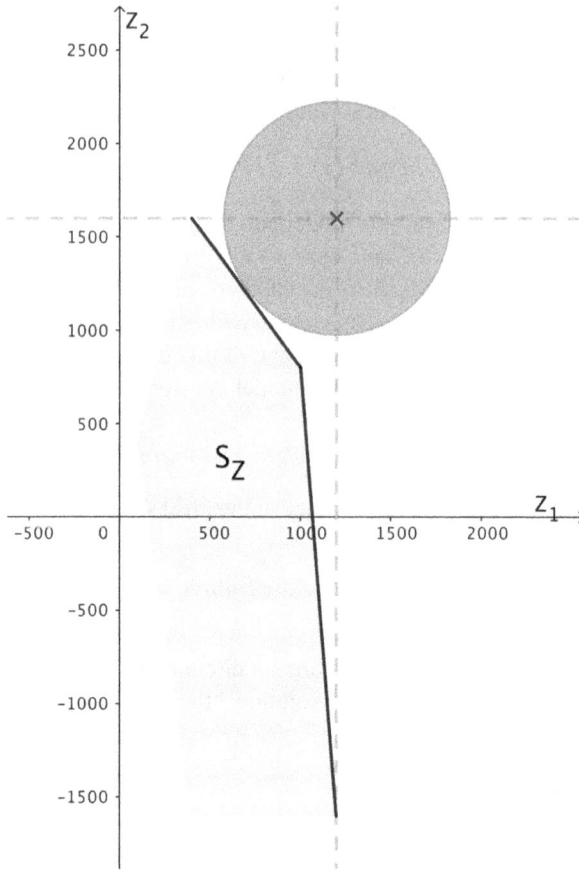

Figure 9.14: Feasible Region on the Criterion Space, Ideal Solution and Compromise Solution that Minimizes the L_2-Distance to the Ideal Solution.

$$\min\left\{\max_{p=1,\dots,k} \lambda_p\left(Z_p^* - Z_p(x)\right) - \sum_{p=1}^{k} p_p Z_p(x)\right\}$$

s.t. $x \in S$

where $\lambda_p > 0$, $p = 1, \dots, k$ and $\sum_{p=1}^{k} \lambda_p = 1$, are the (fixed) weights used to assign a different "importance" to each objective; p_p are sufficiently small positive scalars. For more detail, refer to Antunes et al. (2016).

9.4   Final Remarks

This chapter addressed two very common scalarization methods: the weighted sum and the distance minimization. Some examples highlight some of the methods' weaknesses and propose strategies to overcome them. None of these methods reduces the feasible region since they do not introduce new constraints that cut the feasible region. Only the Tchebychev distance minimization models' adds new constraints. However, these do not act as cuts on the feasible region.

Concerning the moment when the DM's preferences are elicited and aggregated to the model, both methods, as presented in this chapter, may be considered as a *priori* methods. Preferences are modelled by the weights and by the choice of the distance function, which have been set before the method is applied. However, both methods can be used to generate a (large) number of efficient solutions if parameters are systematically varied and the corresponding solution determined. The computed solutions can then be presented to the DM so (s)he can choose the most interesting one. If this latter strategy is adopted both methods are considered to not have any articulation of preferences. The distance minimization model becomes known as the Method of Global Criterion (Miettinen, 2008).

These two methods, although addressed independently in this chapter, have some connection points. Costa and Climaco (1999) showed a close relation between reference point and weights.

9.5 Exercises

9.5.1 Solved Exercises

Consider the MOPL problem with three objective functions presented below. Figure 9.15 shows the feasible region (set S), the objective functions and the corresponding gradient vectors and optimal solutions ($Z_1^*(4, 4) = 12$, $Z_2^*(3, 1) = -2$, $Z_3^*(4, 3) = 9$). It also depicts the Efficient Frontier and the dominance cone.

$$\text{Max } Z_1 = 2x + y$$
$$\text{Min } Z_2 = -x + y$$
$$\text{Max } Z_3 = 3x - y$$
$$\text{s.t. } x + 3y \geq 6$$
$$2x - y \leq 5$$
$$-x + y \leq 4$$
$$x + 2y \leq 12$$
$$x \leq 4$$
$$x, y \geq 0$$

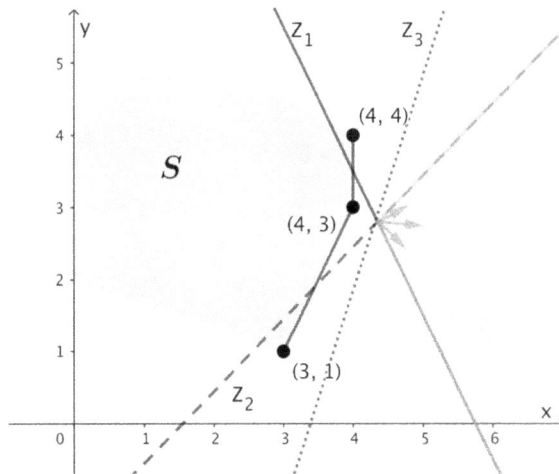

Figure 9.15: Feasible Region, Optimal Solutions, Dominance Cone and Efficient Frontier.

(a) Suppose the DM's preferences concerning the objectives are $Z_3 \gg Z_2 \gg Z_1$. Propose a set of weights reflecting the DM's preferences. Determine the compromise solution one obtains with the weighted sum method.

Following the DM's preferences, a possible weight set can be $\lambda_1 = 0.1$, $\lambda_2 = 0.3$ and $\lambda_3 = 0.6$. However, one of the objectives is a minimization one, therefore, care should be taken since the weighted sum model "assumes" all objectives to have the same optimization directions (they should all be "Max" or all be "Min"). Given the equivalence

$$\text{Min } Z_2 = -x + y \Leftrightarrow -\text{Max } Z_2' = x - y$$

The weighted sum objective function can be written as

$$\text{Max } Z = \lambda_1 Z_1 + \lambda_2 Z_2' + \lambda_3 Z_3.$$

Therefore, the weighted sum model is

$$\text{Max } Z = 2.3x - 0.8y$$
$$\text{s.t.} \quad (x, y) \in S$$

Figure 9.16 shows the weighted sum objective function over the feasible region. The compromise solution is then vertex (4, 3) whose value in the three objective functions are:

$$Z_1 (4, 3) = 11, Z_2 (4, 3) = -1, Z_3 (4, 3) = 9.$$

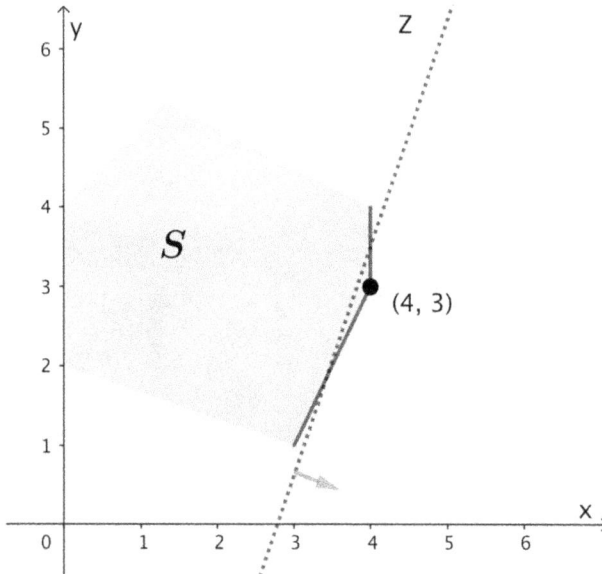

Figure 9.16: Objective Function that maximizes the weighted sum model. Corresponding Gradient Vector and Optimal Solution.

In such a small example where all (individual) optimal solutions are vertices of the efficient frontier and, apart from these, there is no other extreme point in the efficient frontier, the optimal solution of the weighted sum model will always be the optimal vertex of some of the objective functions. In this case, it turned out to be the optimal vertex for Z_3.

If instead of changing the optimization direction of Z_2, one would have changed the direction of the other two objective functions, what would be the differences?

Firstly, let's find the minimization functions that are equivalent to Z_1 and Z_3:

$$\text{Max } Z_1 = 2x + y \Leftrightarrow -\text{Min } Z_1{}' = -2x - y$$

$$\text{Max } Z_3 = 3x - y \Leftrightarrow -\text{Min } Z_3{}' = -3x + y$$

The weighted sum objective function can be written as

$$\text{Min } Z = \lambda_1 Z_1{}' + \lambda_2 Z_2 + \lambda_3 Z_3{}'.$$

Notice the optimization direction of the weighted sum model needs to be in accordance with the objective functions that compose it. The model formulation is then

$$\text{Min } Z = -2.3x + 0.8y \Leftrightarrow -\text{Max } Z = 2.3x - 0.8y.$$

The optimal solution is the same. The only difference is the optimal value of the weighted sum function. The optimal values are opposite in signs.

(b) Formulate the model that minimizes the L_1 distance to the ideal solution. Determine its optimal solution.

The optimal value of each objective function is $Z_1{}^* = 12$, $Z_2{}^* = -2$, $Z_3{}^* = 9$. Therefore the L_1-distance minimization function to the ideal solution is given by

$$\text{Min } |2x + y - 12| + |-x + y - (-2)| + |3x - y - 9|$$
$$\text{s. t. } (x, y) \in S$$

In order to have a linear function, the absolute value function needs to be removed. Remember the optimal values are the highest/lowest an objective function attains over the feasible region. If the optimization direction is maximization, the value is the highest, then $|2x + y - 12| = 12 - (2x + y)$ and $|3x - y - 9| = 9 - (3x - y)$. While for minimization direction, the optimal value is the lowest and so $|-x + y - (-2)| = -x + y + 2$. The linear model formulation that minimizes the L_1 distance to the ideal solution is given below. Figure 9.17 depicts the objective function over the feasible region and the corresponding gradient vector. The optimal solution is vertex (4, 3).

$$\text{Min } Z = 23 - 6x + y$$
$$\text{s.t. } (x, y) \in S$$

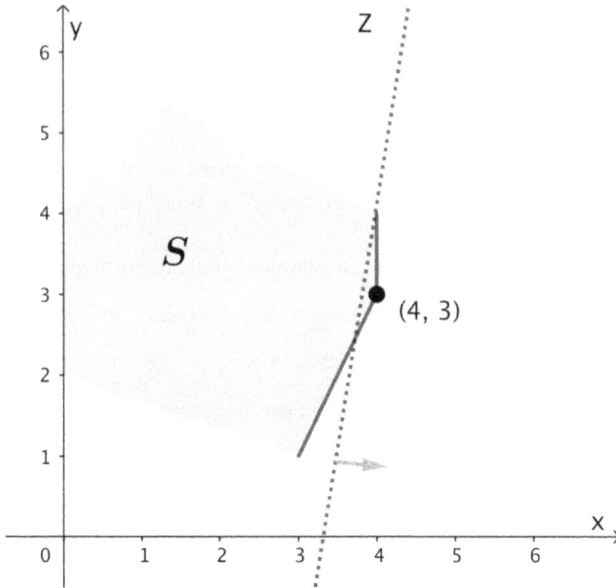

Figure 9.17: Objective Function that Minimizes L_1 Distance to the Ideal Solution. Corresponding Gradient Vector and Optimal Solution.

(c) Formulate the model that minimizes the L_∞ distance to the ideal solution. Determine its optimal solution.

The L_∞-distance minimization function to the ideal solution is given by

$$\text{Min Max}(|2x + y - 12|, |-x + y - (-2)|, |3x - y - 9|)$$
$$\text{s.t. } (x, y) \in S$$

The linear model formulation that minimizes the L_∞ distance to the ideal solution is given below. The optimal solution is again vertex $(4, 3)$ since it is the one that minimizes the maximum value of vertex υ. As mentioned above, the introduction of variable υ makes it impossible to solve the problem graphically and an optimization software is needed to determine the optimal solution. In this small example, however, since the efficient frontier has only three vertices and, hence, one of them is the optimal solution, to determine the optimal solution one needs only to calculate the distance corresponding to each vertex and make the adequate choice.

$$\text{Min } Z = \upsilon$$
$$\text{s.t. } \quad 12 - (2x + y) \geq \upsilon$$
$$-x + y - (-2) \leq \upsilon$$
$$9 - (3x - y) \geq \upsilon$$
$$(x, y) \in S$$

9.5.2 Proposed Exercises

1. Normalize each of the following vectors according to the L_1, L_2 and L_∞-norms:
   (i) $(3, 5)$
   (ii) $(4, 3, 6)$
   (iii) $(2, -3, -1)$
   (iv) $(-1, 1, 0, 1, 0)$
   (v) $(-1, 2, 0, 1)$
   (vi) $(-2, 3, 4, -1, 0)$

2. Apply the L_1, L_2 and L_∞-metrics to calculate the distances between x and y where:
   (i) $x = (2, -1, 0, -3)$ and $y = (1, -2, -1, 5,)$
   (ii) $x = (2, 2, 2)$ and $y = (0, 2, 0)$
   (iii) $x = (3, 1, 0, 2)$ and $y = (2, 0, -1, 3)$

   If Chapter 7's exercises have been solved, the first question of most of the exercises below has already been solved.

3. Consider the MOLP problem

$$\text{Max } Z_1 = -x + 3y$$
$$\text{Max } Z_2 = 3x + 3y$$
$$\text{Max } Z_3 = x + 2y$$
$$\text{s.t.} \qquad y \leq 4$$
$$x + 2y \leq 10$$
$$2x + y \leq 10$$
$$x, y \geq 0$$

   (a) Graph the feasible region in the decision space and determine, which solutions might be of interest to the decision maker.
   (b) Apply the weighted sum method to determine a compromise solution assuming equal importance among the objective functions
   (c) Consider the objectives ordered as $Z_2 \gg Z_1 \gg Z_3$. Propose a weight vector reflecting such ordering. Determine the corresponding compromise solution.

4. Consider the MOLP problem

$$\text{Max } Z_1 = y$$
$$\text{Max } Z_2 = x - 2y$$
$$\text{s.t.} \qquad 5x - 2y \leq 40$$
$$3x + 5y \leq 55$$
$$-x + y \leq 3$$
$$x + y \geq 3$$
$$x, y \geq 0$$

(a) Graph the feasible region and determine the Pareto frontier.
(b) Assume the decision maker considers objective 2 to be the most important. Propose a weight vector that reflects the decision maker's preferences regarding the objective functions. Determine the corresponding compromise solution.
(c) Solve the parametric problem one obtains when maximizing function Z as the following convex combination:

$$Z = \lambda Z_1 + (1 - \lambda) Z_2, \text{ with } \lambda \in [0, 1].$$

5. Consider the MOLP problem

$$\text{Max } Z_1 = 210x + 500y$$
$$\text{Max } Z_2 = -x + 2y$$
$$\text{Max } Z_3 = 0.1x - 0.6y$$
$$\text{s.t.} \qquad 3x + y \geq 7$$
$$-x + 2y \leq 7$$
$$x + 2y \leq 17$$
$$-x + 2y \geq 0$$
$$2x - 2y \leq 9$$
$$x, y \geq 0$$

Determine the compromise solution assuming the weight vector $\Lambda = (0.2, 0.3, 0.5)$. Re-scale the objective functions using:
(a) L_1-metric
(b) L_2-metric
(c) L_∞-metric
(d) An adequate power of 10
(e) The range equalization factor.

6. Consider the MOLP problem

$$\text{Max } Z = (2x_1 + x_2; -x_1 + x_2)$$
$$\text{s.t.} \qquad x_1 + x_2 \geq 2$$
$$2x_1 - x_2 \leq 4$$
$$-x_1 + x_2 \leq 4$$
$$x_1 + 2x_2 \leq 12$$
$$x_1, x_2 \geq 0$$

(a) Solve each objective function independently and determine the efficient frontier.
(b) Solve the weighted sum problem in the objectives space.
(c) Propose a model formulation and find the compromise solution that is closer to the ideal solution when applying the
    (i) L_1-distance
    (ii) L_∞-distance

(d) Consider the new objective function Min $Z_3 = -2x_1 + 2x_2$.
   (i) Determine the new efficient frontier and the ideal solution
   (ii) Assuming $Z_1 \gg Z_2$ and $Z_1 \gg Z_3$, propose a weight vector Λ that models such relations. Solve the corresponding weighted sum model.
   (iii) Solve the problem that minimizes the L_1-distance to the ideal solution.
   (iv) Solve the problem that minimizes the L_∞-distance to the ideal solution.

7. Consider the MOLP problem

$$\text{Max } Z = (2x_1 + x_2; x_1 - x_2)$$
$$\text{s.t.} \quad x_1 + x_2 \geq 2$$
$$-2x_1 + x_2 \geq 0$$
$$-x_1 + x_2 \leq 4$$
$$x_1, x_2 \geq 0$$

(a) Determine the Pareto Front.
(b) Solve the weighted sum problem in the objectives space. Propose a compromise solution that should be the most satisfying one to the DM. Justify your choice.
(c) Find the compromise solution that is closer to the ideal solution when applying the
   (i) L_1-distance
   (ii) L_∞-distance
(d) Propose a model formulation for the problem that minimizes the L_2-distance to the ideal solution.

8. Consider the MOLP problem

$$\text{Max } Z_1 = 2x + y$$
$$\text{Min } Z_2 = 0.5x + y$$
$$\text{s.t.} \quad x \geq 3$$
$$y \geq 4$$
$$x \leq 8$$
$$-x + y \geq 3$$
$$0.5x + y \leq 12$$

(a) Solve each objective function independently and determine the efficient frontier.
(b) Solve the weighted sum problem considering both objectives have the same importance.
(c) Determine all compromise solutions one can obtain when applying the weighted sum model. For each solution, determine the corresponding weight intervals.

(d) Find the compromise solution that is closer to the ideal solution when applying the
  (i) L_1-distance
  (ii) L_∞-distance
(e) Propose a model formulation for the problem that minimizes the L_2-distance to the ideal solution.

9. Consider the MOLP problem

$$\text{Max } Z_1 = 500x + 200y + 300z$$
$$\text{Max } Z_2 = 2x - y + z$$
$$\text{Max } Z_3 = -10x - 20y + 40z$$
$$\text{s.t.} \qquad 1 \le x \le 2$$
$$1 \le y \le 2$$
$$1 \le z \le 2$$

Knowing that $Z_1 \in [1000, 2000]$, $Z_2 \in [1, 5]$ and $Z_3 \in [-20, 50]$, determine the compromise solution when the objective functions have all the same importance.

10. Aunt Leonarda recently bought 22 kilograms of cregola, a very rare substance used in perfumery. She wants to use it to produce her two most famous perfumes, named as X and Y, for reasons of confidentiality. Aunt Leonarda wonders: "Each kiloliter of X gives me a profit of one thousand pilins, while with the perfume Y I manage to collect profits of 3 thousand pilins per kiloliter. But, each kiloliter of Y consumes two kilos of cregola, and with X the consumption is one to one...."

"Don't forget that to fulfil the production plan, at least three kiloliters of X and two of Y should be produced, ... and that our distributor can only assure selling eight kiloliters of Y" ,- added Blimunda, Aunt's secretary. – "And there is something else, ... the total of X produced has to exceed the total production of Y by six kiloliters" ,- Blimunda reminded her.

- "What a mess,... I want to minimize the consumption of cregola with this production, but I also want to make the highest profit." Admitted the friendly Aunt Leonarda.

Blimunda wrote something on paper and proudly presented the following formulation to Aunt Leonarda.

$$\text{Max } Z(-x_1 - x_2; x_1 + 3x_2)$$
$$\text{s.t.} \qquad x_1 \ge 3$$
$$x_2 \ge 2$$
$$x_2 \le 8$$
$$x_1 - x_2 \le 6$$
$$x_1 + 2x_2 \le 21$$

Aunt Leonarda cannot decide which objective is the most important.

(a) Comment on the above formulation, establishing the relation between each constraint/objective function with Aunt's and Blimunda's observations.

Solve the problems below using the proposed formulation.

(b) Suppose Aunt Leonarda gives the same weight to both objectives. Find the corresponding compromise solution.

(c) Discuss the compromise solution using the weighted sum model.

(d) Find the compromise solution that is closer to the ideal solution when applying the
   (i) L_1-distance
   (ii) L_∞-distance

(e) Propose a model formulation for the problem that minimizes the L_2-distance to the ideal solution.

11. Jaime is preparing his study time for next week. He will have two tests (Algebra and Biochemistry). For each study hour of Algebra his girlfriend will offer him 5 dl of strawberry ice cream. But she already threatened him that she would reduce 2 dl of ice cream for every hour he studied Biochemistry. Note that his girlfriend is a Math student... Every hour of Biochemistry study will give him 4 units of satisfaction. But an hour of Algebra study will provide him with a unit of negative satisfaction ...

Jaime does not want to study more than 6 hours of Algebra, but the amount of Biochemistry study hours should not exceed in 3 hours the Algebra study period. He would like to maximize the amount of ice cream he will receive, but also the total satisfaction with the study.

This problem can be formulated as

$$\text{Max } Z_1 = 5x - 2y$$
$$\text{Max } Z_2 = -x + 4y$$
$$\text{s.t.} \quad -x + y \leq 3$$
$$x \leq 6$$
$$x, y \geq 0$$

(a) Solve each objective function independently and determine the Pareto Front.

(b) Knowing that ice cream is the most important objective for Jaime, determine the set of weights that provide, as compromise solution, the optimal solution of ice cream objective function.

(c) Find the compromise solution that is closer to the ideal solution when applying the
   (i) L_1-distance
   (ii) L_∞-distance

(d) Propose a model formulation for the problem that minimizes the L_2-distance to the ideal solution.

12. In a factory there are two types of machines used in the production lines of four products. The available capacities of each type of machine (in hours / week) and the number of processing hours/week required at each machine, per unit produced product, are presented in the table:

Machine	P1	P2	P3	P4	Availability
A	2	1	4	3	60
B	3	4	1	2	60

The company management wants to maximize profit, quality of P1, P2, P3 and P4 supply and the degree of employee satisfaction. Recent studies have made it possible to establish the unit contribution of each product to each objective:

	P1	P2	P3	P4
Profit	3	1	2	1
Quality	1	-1	2	4
Satisfaction	-1	5	1	2

(a) Formulate the problem as a multi-objective linear programming model.
(b) Below, the objectives are sorted in decreasing order of importance. Propose a weight vector that models the DM preferences. Using an optimization software (e.g. GAMS), determine the compromise solution that corresponds to each one of them.
   (i) satisfaction, profit, quality,
   (ii) quality, satisfaction, profit,
   (iii) quality, profit, satisfaction,
   (iv) profit, quality, satisfaction.
(c) Varying the weight vector, use the weighted sum method to approximate the Pareto Front.
(d) Find the compromise solution that is closer to the ideal solution when applying the
   (i) L_1-distance
   (ii) L_∞-distance

References

Antunes, C.H., M.J. Alves and J. Climaco (2016). *Multiobjective Linear and Integer Programming*. Springer, New York, NY.
Cohon, J.L. (2003). *Multiobjective Programming and Planning*. Dover Publication Inc.
Costa, J.P. and J.C. Clímaco (1999). Relating reference points and weights in MOLP. *Journal of Multi-Criteria Decision Analysis*, 8(5), 281–290.
Gass, S. and T. Saaty (1955). The computational algorithm for the parametric objective function. *Naval Research Logistics Quarterly*, 2(1-2), 39–45.

Miettinen, K. (1998). *Nonlinear Multiobjective Optimization* (Vol. 12). Springer Science & Business Media.

Miettinen, K. (2008). Introduction to multiobjective optimization: Noninteractive approaches. *In:* Branke, J., K. Deb, K. Miettinen and R. Słowiński (Eds.), Multiobjective Optimization: Interactive and Evolutionary Approaches. Springer, Berlin, Heidelberg.

Steuer, R.E. (1986). *Multiple Criteria Optimization. Theory, Computation and Applications*. John Wiley and Sons, Inc.

Winston, W.L. (2002). *Introduction to Mathematical Programming: Applications and Algorithms*. Duxbury.

Interactive Methods

10.1 Introduction

Since the 1970's, the development of interactive methods to solve multiobjective optimization models has been an active area of research (Korhonen 2005). With these methods, the decision maker (DM) provides the preference information during an iterative process. With the given information, the methods progress in the search of the compromise solution. Hence, the DM plays a very important and central role during the process of reaching the final solution. (S)he have to be available and willing to actively participate in the process by answering questions (some of them hard) that will reveal her/his preference system. Otherwise, a different kind of method has to be chosen. The preference system does not need to be clear for the DM. The dialog phase between the DM and the analyst, present in all interactive methods, allows the DM to get acquainted with their preferences. In the end, the DM not only has a compromise solution for the problem but (s)he have a deeper understanding on what they value.

In general, the interactive methods follow four main steps:
1. Initialization (e.g, to compute reference points, as the ideal and nadir points, and have them validated by the DM).
2. Calculation phase: compute a solution (most methods compute an efficient solution).
3. Dialogue phase: ask the DM to comment on the previous solution to assess how much the solution meets their preferences. If it does, end the process. If not, further questions are asked to understand what needs to be "improved".
4. Incorporate the information given by the DM at the previous step and generate a new solution (i.e., go back to step 2).

The most common stopping criteria is the DM's satisfaction with the solution (Miettinen and Hakanen 2017). Nonetheless, some interactive methods may propose special stopping criteria (e.g. a preset number of iterations).

This kind of approach to solve multiobjective problems was first proposed by Benayoun et al. (1971) with STEM methods (which stands for the STEP Method). This chapter will focus mainly on this method given its pioneering role in the multiobjective interactive methods (Tamiz and Jones 1997). Moreover, the method builds upon the concepts of distance minimization presented in Chapter 9, and of reduced feasible region as studied in Chapter 8. Lastly, STEM is a simple method which makes it particularly suitable for educational purposes.

Generically, a multiobjective linear programming (MOLP) model can be written on the form:

$$\text{Max } Z_1 = \sum_{i=1}^{n} c_{i1} x_i$$

$$\text{Max } Z_2 = \sum_{i=1}^{n} c_{i2} x_i$$

$$...$$

$$\text{Max } Z_k = \sum_{i=1}^{n} c_{ik} x_i$$

$$\text{s.t.} \sum_{i=1}^{n} A_{ij} x_i \geq b_j, \quad \forall j = 1, ..., m$$

$$x_i \geq 0, \quad \forall i = 1, ..., n$$

where c_{ip}, $i = 1, ..., n$, $p = 1, ..., k$ are the coefficients of the objective functions, A_{ij} is the $n \cdot m$ matrix with constraints coefficients and b_j, $j = 1, ..., m$ are the constant terms on the right side of the constraints. This problem is comprised of n variables, m constraints and k objective functions[1].

Example 10.1 will be solved alongside the methods' presentation.

Example 10.1

The DM aims at determining the compromise solution of three objective function problems presented below. Figure 10.1 depicts the feasible region (set S), the three objective functions and their gradient vectors.

The optimal solutions of each objective function when optimized independently[2] are

$$x_1^* = (40, 40) \text{ with } Z_1^* = 1200,$$
$$x_2^* = (0, 40) \text{ with } Z_2^* = 1600,$$
$$x_3^* = (20, 60) \text{ with } Z_3^* = 60.$$

---

[1] For brevity, some fundamental concepts on MOLP addressed in Chapter 7 will not be defined in this chapter. However, a supplementary chapter is available for free with all the main definitions and concepts needed to fully understand the topics addressed in this chapter.

[2] For further details on how to graphically solve a linear programming problem, the interested reader should refer to any Operations Research reference book as, for example, Winston (2002).

Max $Z_1 = 20x + 10y$

Max $Z_2 = -80x + 40y$

Max $Z_3 = y$

s.t. $x + y \geq 20$

$2x - y \leq 40$

$-x + y \leq 80$

$x + y \leq 80$

$x, y \geq 0$

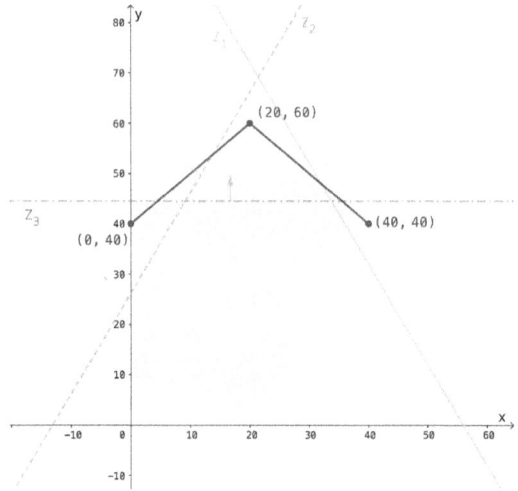

Figure 10.1: Feasible Region and Optimal Solution of Each Objective Function

This chapter is organized into two parts. The first one addresses the STEM method in detail (Section 10.2). Its subsections will focus on the augmented weighted Tchebychev distance function, which generalizes the distance function addressed in Chapter 9; the presentation of the STEM iterative process where two examples will be solved to highlight different features; and as final remarks some of these methods' limitations will be addressed. The second part of this chapter briefly presents three other interactive methods (Section 10.3): two very well known as the Zionts-Wallenius and the Pareto Race methods, and a more recent one, the NATILUS. This chapter ends with some final remarks (Section 10.4) and a set of exercises addressing the topics covered (Section 10.5).

10.2 The Step Method (STEM)

The step method (STEM) is one of the first interactive methods introduced for multiobjective optimization. It was originally developed for MOLP problems (Benayoun et al. 1971) and later on extended to integer and nonlinear multiobjective problems. In its first version, it would converge to the best compromise solution in no more than k steps (where k is the number of objective functions). This was very important at the time given the computational limitation existing in the 1970's. The method is based on distance minimization to the ideal solution with some adaptations to integrate the information provided by the DM at the dialogue phase. This method successively reduces the feasible region using this preferences information.

Before entering into the method's details, one needs to extend Tchebychev distance function definition with respect to the one provided in Chapter 9.

10.2.1 Augmented Weighted Tchebychev Distance

At the calculation phase, the STEM method minimizes the Tchebychev distance to a reference point. This metric assures the minimization of the distance for the objective with the largest distance to the reference point.

Any distance function can be minimized in its simplest form as weighted or as weighted and augmented functions. For the Tchebychev distance, these three functions are:

$$\text{distance: } L_\infty = \|x - y\|_\infty = \max_{i=1,\dots n} |x_i - y_i|$$

$$\text{weighted distance: } L_\infty^\lambda = \|x - y\|_\infty^\lambda = \max_{i=1,\dots n} \lambda_i |x_i - y_i|$$

augmented weighted distance:

$$L_\infty^{\lambda\varepsilon} = \|x - y\|_\infty^{\lambda\varepsilon} = \max_{i=1,\dots n} \lambda_i |x_i - y_i| + \sum_{i=1}^{n} \varepsilon_i |x_i - y_i|$$

where $x, y \in R^n$, $\sum_{i=1}^{n} \lambda_i = 1$, $\lambda_i > 0$ and $\varepsilon_i \geq 0$, $i = 1, \dots, n$.

Figure 10.2 shows the contour line for the equidistant points to the origin when considering the weighted Tchebychev distance (L_∞^λ) and the augmented weighted Tchebychev distance ($L_\infty^{\lambda\varepsilon}$). It can be shown that the augmented

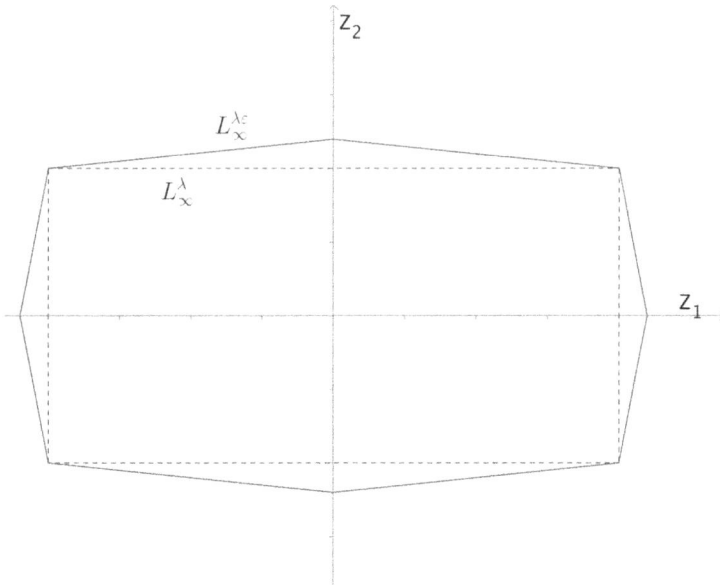

Figure 10.2: Contour Line for Points Equidistant from the Origin for L_∞^λ and $L_\infty^{\lambda\varepsilon}$

weighted Tchebychev distance minimization problem produces only efficient solutions (Antunes et al., 2016).

10.2.2 Iterative Process

The STEM iterative process has seven steps. The three first steps define the method initialization procedure. Steps 4 and 5 are the calculation phase. The dialogue phase is addressed in step 6. The last step incorporates the information given by the DM and sets the parameters for the next iteration.

Step 1: Compute the payoff matrix. This matrix is built by optimizing each of the k objective functions individual and computing the value of the remaining $k - 1$ functions at the optimal solution[3]. Set iteration counter $h = 0$.

Step 2: From the payoff matrix, determine the ideal solution Z^* and the nadir solution N^* such that

$$Z^* = (z_1^*, ..., z_k^*) \text{ and } N^* = (n_1^*, ..., n_k^*)$$

where $z_p^* = \max_{j=1,...,k} z_j^p$, $n_p^* = \min_{j=1,...,k} z_j^p$ and z_j^p the value of row j and column i in the payoff matrix.

Step 3: Compute the weights β_p, $p = 1, ..., k$, given by

$$\beta_p = \begin{cases} \dfrac{z_p^* - n_p^*}{z_p^*}\left[\sum_{i=1}^{n} c_{ip}^2\right]^{-\frac{1}{2}}, & \text{if } z_p^* \geq 0 \\[4mm] \dfrac{n_p^* - z_p^*}{n_p^*}\left[\sum_{i=1}^{n} c_{ip}^2\right]^{-\frac{1}{2}}, & \text{if } z_p^* < 0 \end{cases} \qquad [10.1]$$

These β_p values are weights that take into account the range of the objective function values over a set of non-dominated solutions (factor on the left of the expression) and the scale of the objective functions (the second factor of the expression). This latter factor is a normalization term according to the Euclidean norm. These values will define the weights at the calculation phase, where the weighted Tchebychev distance to the ideal solution is minimized.

Step 4: Let I^* be the set of the objective functions that are to be relaxed in the next iteration (at the first iteration, $I^* = \varnothing$ since no objective has yet been relaxed). Let, $h = h + 1$. At the current iteration, iteration h, the weights to be used in the weighted Tchebychev distance are

$$\alpha_p^{(h)} = \begin{cases} 0 & , \text{if } p \in I^* \\[3mm] \dfrac{\beta_p}{\sum_{p \in I^*} \beta_p} & , \text{if } p \notin I^* \end{cases} \qquad [10.2]$$

---
[3] The payoff table has been addressed with detail in Chapter 7.

Note that in all iterations, the $\alpha_p^{(h)}$ sum to one.

Step 5: Solve the weighted Tchebychev distance minimization to the ideal solution model given by

$$\text{Min } \upsilon$$
$$\text{s.t. } \upsilon \geq \alpha_p^{(h)}\left(z_p^* - Z_p(x)\right), \ p = 1, \ldots, k$$
$$x \in S^{(h)}$$
$$\upsilon \geq 0$$

The compromise solution in the objectives space determined by the above model will be $z^{(h)}$, which corresponds to the efficient solution $x^{(h)}$ from the reduced feasible region $S^{(h)}$. Note that $S^{(1)} = S$ where S is the feasible region. This means the algorithm starts considering the initial feasible region (not yet reduced).

Step 6: Compare $z^{(h)}$ with Z^*. This comparison aims at providing the DM with a way to evaluate if $z^{(h)}$ is of "good quality" when compared to the ideal solution. If all the components of $z^{(h)}$ (the values of $Z_p(x^{(h)})$, $p = 1, \ldots, k$) are satisfactory to the DM, then STOP. Otherwise, ask the DM to indicate which objective function (s)he is willing to sacrifice and how much will (s)he allow it to reduce (Δ_p) from that objective value. Go to Step 7.

Step 7: Assume the DM has chosen objective p, with $p \notin I^*$ as the one to be reduced by Δ_p. Update set I^*, $I^* = I^* \cup \{p\}$, and the reduced feasible region for the next iteration $(h + 1)$ such that

$$S^{(h+1)} = \left\{x \in S^{(h)} : Z_p(x) \geq z_p^{(h)} - \Delta_p, \ p \in I^* \text{ and } Z_p(x) \geq z_p^{(h)}, \ p \notin I^*\right\} \quad [10.3]$$

Return to step 4.

Figure 10.3 displays the iterative process described above.

Let's apply the STEM method to Example 10.1 considering only the two first objective functions so one can see how the procedure evolves in the objectives space.

Example 10.2

STEM first steps are the building of the payoff matrix (Table 10.1) and the computation of the ideal and the nadir points: $Z^* = (1200, 1600)$ and $N^* = (400, -1600)$. The model is presented below for reference.

$$\text{Max } Z_1 = 20x + 10y$$
$$\text{Max } Z_2 = -80x + 40y$$
$$\text{s.t.} \quad x + y \geq 20$$
$$2x - y \leq 40$$
$$-x + y \leq 40$$
$$x + y \leq 80$$
$$x, y \geq 0$$

start

Built payoff table

Determine the nadir point coordinates
$n_p^*, p = 1, \ldots, k$ and the ideal point
coordinates $z_{p,}^* \, p = 1, \ldots, k$

$S^{(1)} = S$ original feasible region
Compute $\beta_p, p = 1, \ldots, k$

$I^* = \emptyset$ and $h = 1$

Compute $\alpha_p^{(h)}, p = 1, \ldots, k$

Find the efficient solution $\boldsymbol{x}^{(h)}$ and the
corresponding non-dominated solution $\boldsymbol{z}^{(h)}$
in feasible region $\boldsymbol{S}^{(h)}$

$h = h + 1$

$$S^{(h+1)} = \left\{ x \in S^{(h)} \,\middle|\, \begin{array}{l} Z_p(x) \geq z_p^{(h)} - \Delta_p, p \in I^* \\ Z_p(x) \geq z_p^{(h)}, p \notin I^* \end{array} \right\}$$

$I^* = I^* \cup \{p\}$

Is $\boldsymbol{z}^{(h)}$
satisfactory for
the DM?

N → Let $p \notin I^*$ be the objective
to be reduced by Δ_p

S

$\boldsymbol{z}^{(h)}$ is the
compromise solution

end

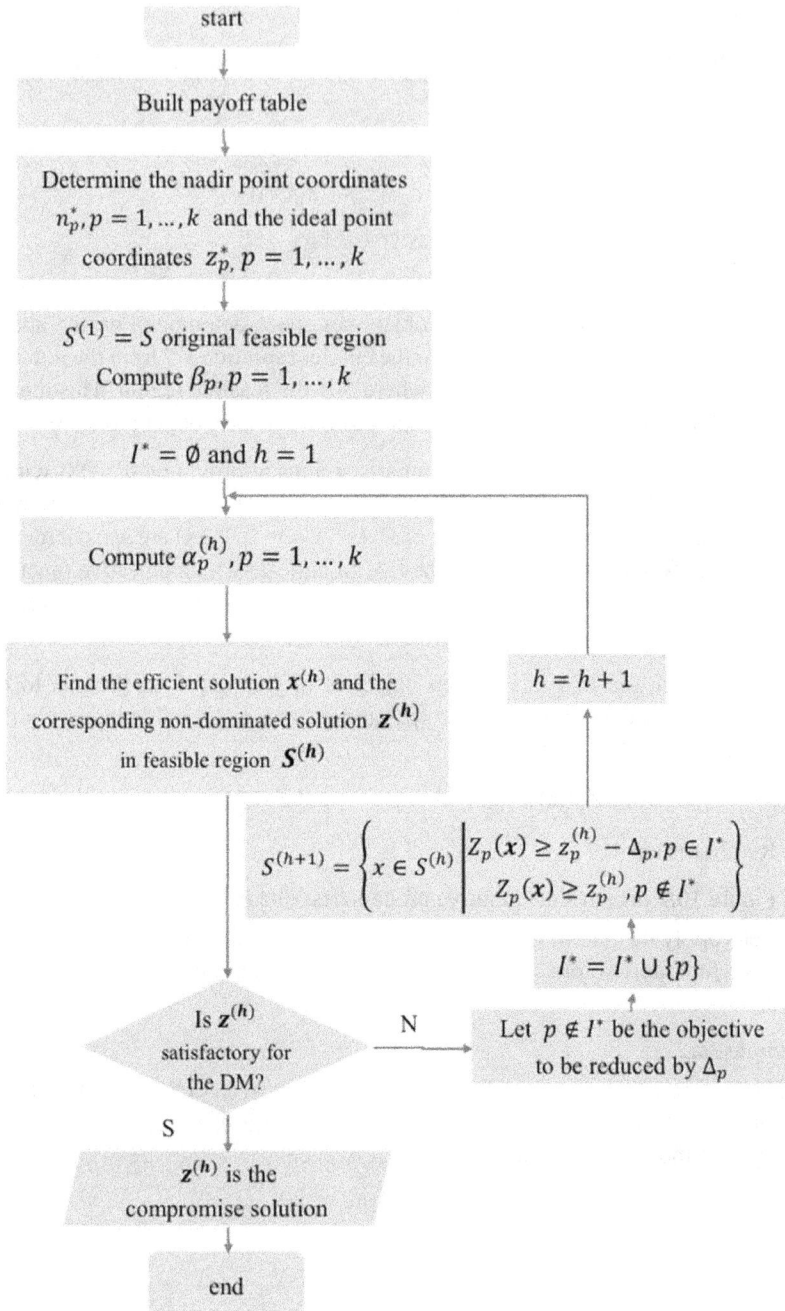

Figure 10.3: STEM Method

Table 10.1: Payoff Matrix

	Z_1	Z_2
$x_1{}^* = (40, 40)$	1200	-1600
$x_2{}^* = (0, 40)$	400	1600

The next step is to compute the β_p given in expression [10.1]. Since all components of the ideal solution are positive, one should compute them as

$$\beta_p = \frac{z_p^* - n_p^*}{z_p^*}\left[\sum_{i=1}^{2} c_{ip}^2\right]^{-\frac{1}{2}}, \; p = 1, 2.$$

Hence,

$$\beta_1 = \frac{z_1^* - n_1^*}{z_1^*}\left[\sum_{i=1}^{2} c_{i1}^2\right]^{-\frac{1}{2}} = \frac{1200 - 400}{1200} \frac{1}{\sqrt{20^2 + 10^2}} = 0.029814$$

$$\beta_2 = \frac{z_2^* - n_2^*}{z_2^*}\left[\sum_{i=1}^{2} c_{i2}^2\right]^{-\frac{1}{2}} = \frac{1600 - (-1600)}{1600} \frac{1}{\sqrt{(-80)^2 + 40^2}} = 0.022361.$$

For the first iteration $I^* = \varnothing$ and $h = 1$. The corresponding α-values are

$$\alpha_1^{(1)} = \frac{\beta_1}{\beta_1 + \beta_2} = 0.571 \text{ and } \alpha_2^{(1)} = \frac{\beta_2}{\beta_1 + \beta_2} = 0.429$$

The model to be solved at this moment is

$$\begin{aligned}
\text{Min } Z &= \upsilon \\
\text{s.t.} \quad &\upsilon \geq 0.571[1200 - (10x + 10y)] \\
&\upsilon \geq 0.429[1600 - (-80x + 40y)] \\
&\upsilon \geq 0 \\
&x, y \in S^{(1)}
\end{aligned}$$

or[4]

$$\begin{aligned}
\text{Min } Z &= \upsilon \\
\text{s.t.} \quad &x + y \geq 20 \\
&2x - y \leq 40 \\
&-x + y \leq 40 \\
&x + y \leq 80 \\
&\upsilon \geq 0.571[1200 - (20x + 10y)] \\
&\upsilon \geq 0.429[1600 - (-80x + 40y)] \\
&x, y, \upsilon \geq 0
\end{aligned}$$

---

[4] The first four constraints of the model below are from the original feasible region. The other two are the weighted Tchebychev distance constraints.

The optimal solution[5] is $x^{(1)} = (13.32, 53.32)$ which corresponds, in the solution space, to the point $z^{(1)} = (799.65, 1067.13)$. Figure 10.4 depicts the feasible region of the first iteration in the objectives space, which is the same as of the original problem since no objective as yet has been relaxed. It also shows the compromise solution $z^{(1)}$ and how the objective function weights relate to each other.

Before proceeding to the second iteration, the DM has to evaluate this solution and assess its quality (dialog phase). Suppose the DM says (s)he wishes the value of Z_2 to be higher. Also, (s)he are comfortable with relaxing Z_1 up to 100 units, so that (hopefully) Z_2 presents a better value. So, $I^* = \{1\}$ and $\Delta_1 = 100$. Increase the iteration counter in one unit, $h = 2$. The weights need to be (re)computed. Since now only Z_2 has not been relaxed, then only $\alpha_2{}^{(2)}$ needs to be (re)calculated. As the weights need to sum to one, then $\alpha_2{}^{(2)} = 1$. The feasible region for the second iteration, given by expression [10.3], is then

$$S^{(2)} = \{x \in S^{(1)}: Z_1(x) \geq z_1{}^{(1)} - \Delta_1 \text{ and } Z_2(x) \geq z_2{}^{(1)}\}.$$

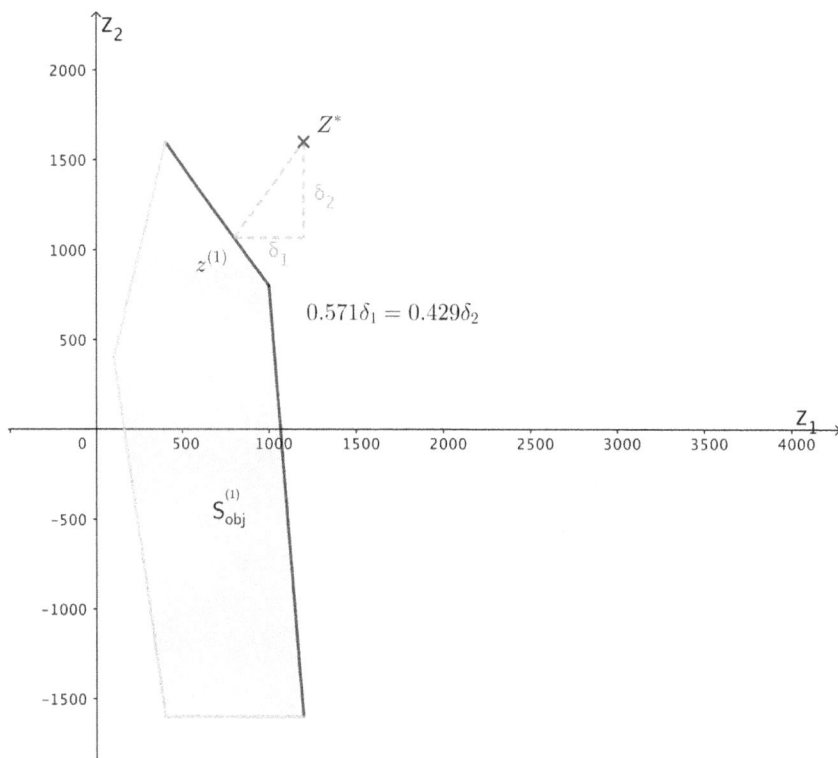

Figure 10.4: Feasible Region in the Objectives Space ($S_Z{}^{(1)}$) and Compromise Solution of STEM First Iteration, $z^{(1)}$.

---

[5] The model in GAMS is available in Appendix A.

The model to be solve in the second iteration is

$$\text{Min } Z = \upsilon$$

s.t. $\quad \upsilon \geq 1600 - (-80x + 40y)$

$$\upsilon \geq 0$$

$$x, y \in S^{(2)}$$

which is equivalent to the model

$$\text{Max } Z_2 = -80x + 40y$$

s.t. $\quad x + y \geq 20$

$$2x - y \leq 40$$

$$-x + y \leq 40$$

$$x + y \leq 80$$

$$20x + 10y \geq 699.65$$

$$x, y, \upsilon \geq 0$$

The optimal solution is $x^{(2)} = (9.9883, 49.9883)$ which corresponds to the non-dominated solution $z^{(2)} = (699.65, 1200.47)$. Fig. 10.5 depicts the feasible region of the second iteration ($S^{(2)}$), which is reduced when compared to $S^{(1)}$. It

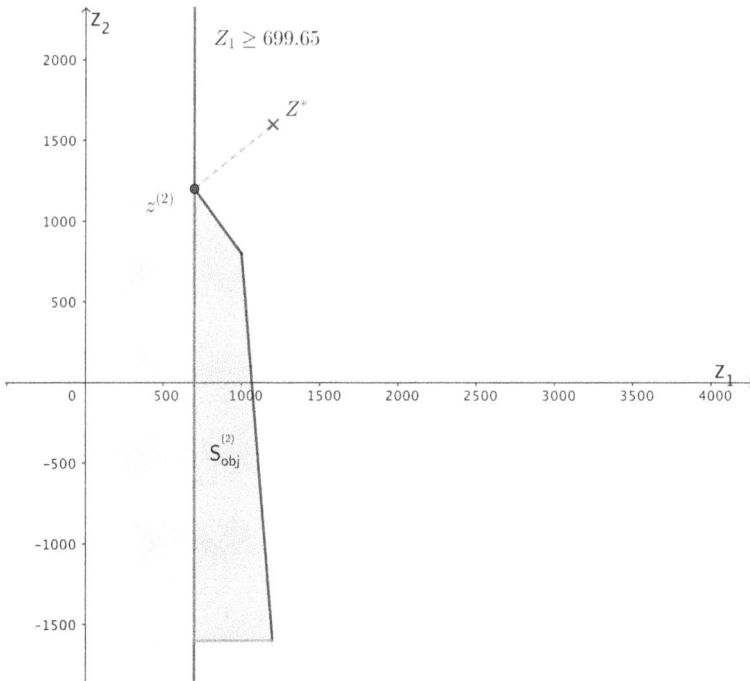

Figure 10.5: Feasible Region ($S_Z^{(2)}$) and Compromise Solution of STEM Second Iteration, $z^{(2)}$.

also shows the compromise solution $z^{(2)}$. According to the first version of the STEM method proposed by Benayoun et al. (1971), the iterative process stops when the number of iterations equals the number of objective functions (in this example, when $h = 2$). Subsequent versions of the method allow the DM to evaluate this solution assessing its quality. When the solution does not meet the DM's preferences, (s)he can reevaluate Δ_1 and determine a new compromise solution. In this case, the STEM method will only finish when the DM is satisfied with the compromise solution.

In the previous example, the objectives' space can be graphically represented since the problem has two objective functions. However, this iterative process is more adequate (interesting) for problems with a larger number of objectives. In Example 10.3, the three objective function problems in Example 10.1 will be solved.

Example 10.3

The first step (since one has already determined the optimal solution for each objective function independently) is to complete the payoff matrix (Table 10.2) and to compute the ideal and nadir points, Z^* and N^*:

$$Z^* = (1200, 1600, 60) \text{ and } N^* = (400, -1600, 40).$$

Table 10.2: Payoff Matrix.

	Z_1	Z_2	Z_3
$x_1^* = (40, 40)$	1200	-1600	40
$x_2^* = (0, 40)$	400	1600	40
$x_3^* = (20, 60)$	1000	800	60

Next the β values need to be calculated and then the three α values for the first iteration (using expressions [10.1] and [10.2], respectively):

$$\beta_1 = 0.007454, \beta_2 = 0.022361, \text{ and } \beta_3 = 0.003727,$$
$$\alpha_1^{(1)} = 0.2222, \alpha_2^{(1)} = 0.6667, \text{ and } \alpha_3^{(1)} = 0.1111.$$

The model to be solved in the first iteration is

$$\text{Min } Z = \upsilon$$

s.t.
$$x + y \geq 20$$
$$2x - y \leq 40$$
$$-x + y \leq 40$$
$$x + y \leq 80$$
$$\upsilon \geq \alpha_1^{(1)} [1200 - (20x + 10y)]$$
$$\upsilon \geq \alpha_2^{(1)} [1600 - (-80x + 40y)]$$
$$\upsilon \geq \alpha_3^{(1)} (60 - y)$$
$$x, y, \upsilon \geq 0$$

The optimal solution[6] is $x^{(1)}$ = (5.33, 45.33) which corresponds, in the objectives space, to the point $z^{(1)}$ = (559.9, 1386.8, 45.33). Figure 10.6 depicts the feasible region in the decision space for the first iteration (the original feasible region since no cut has yet been added).

The DM is not comfortable with this solution and is willing to relax Z_2 by 50% since Z_2 is the least important objective amongst the three.

(2nd iteration) Let $I^* = \{2\}$ and $\alpha_2^{(2)} = 0$. The new α values are:

$$\alpha_1^{(2)} = 0.6667 \text{ and } \alpha_3^{(2)} = 0.3333$$

$$\text{Min } Z = \upsilon$$

$$\text{s.t.} \quad x + y \geq 20$$
$$2x - y \leq 40$$
$$-x + y \leq 40$$
$$x + y \leq 80$$
$$-80x + 40y \geq 0.45 * 1386.8$$
$$20x + 10y \geq 559.9$$
$$y \geq 45.33$$
$$\upsilon \geq \alpha_1^{(2)} [1200 - (20x + 10y)]$$
$$\upsilon \geq \alpha_3^{(2)} (60 - y)$$
$$x, y, \upsilon \geq 0$$

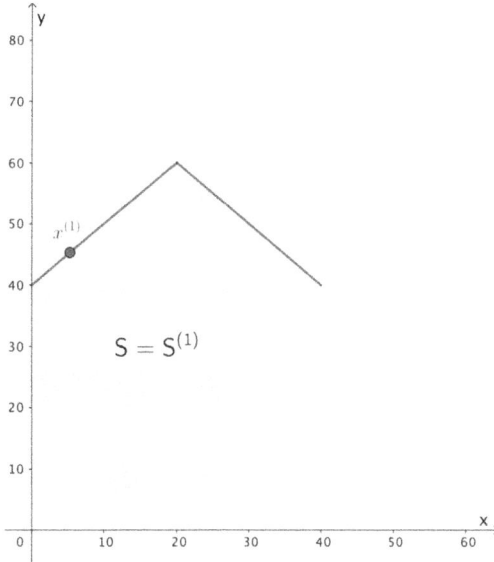

Figure 10.6: Feasible Region in the Decision Space ($S^{(1)}$) and Compromise Solution of STEM First Iteration, $x^{(1)}$.

---

[6] The model in GAMS is available in Appendix A.

The optimal solution is $x^{(2)} = (20.88, 59.11)$ which corresponds, in the objectives space, to the point $z^{(2)} = (1008.7, 694, 59.11)$. Figure 10.7 depicts the feasible region in the decision space for the second iteration, $S^{(2)}$. In this iteration, the feasible region has been considerably reduced when compared with the previous iteration.

After presenting and enquiring once again with the DM about the satisfiability of this solution, (s)he says that it has improved but (s)he wished the value of Z_1 to be further improved and that (s)he doesn't mind relaxing the value of Z_3 by 15%.

(3rd iteration) Let then $I^* = \{2,3\}$ and $\alpha_2^{(3)} = \alpha_3^{(3)} = 0$. Therefore, $\alpha_1^{(3)} = 1$. The model for this iteration is given below. Values in bold are the ones that have changed from the previous iteration.

$$\text{Min } Z = \upsilon$$

$$\text{s.t.} \qquad x + y \geq 20$$
$$2x - y \leq 40$$
$$-x + y \leq 40$$
$$x + y \leq 80$$
$$-80x + 40y \geq 0.45 * 1386.8$$
$$y \geq \mathbf{0.85 * 59.11}$$
$$20x + 10y \geq \mathbf{1008.7}$$
$$\upsilon \geq \alpha_1^{(3)} [1200 - (20x + 10y)]$$
$$x, y, \upsilon \geq 0$$

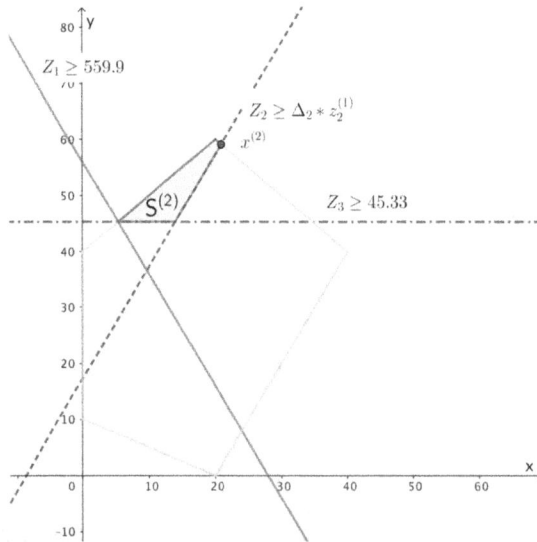

Figure 10.7: Feasible Region in the Decision Space ($S^{(2)}$) and Compromise Solution of STEM Second Iteration, $x^{(2)}$

The optimal solution is $x^{(3)}$ = (20.88, 59.11). This means that the solution didn't change although the value of Z_3 has been relaxed. Fig. 10.8 is a closeup of the feasible region. One sees that $S^{(3)}$ is reduced to a single point ($x^{(2)}$). In cases like this, either the DM accepts $x^{(2)}$ as the compromise solution for the problem or (s)he revises the Δ-values.

10.2.3 Final Remarks

One of the difficulties of this method is to estimate Δ_j, i.e, how much objective *i* should be relaxed in order to potentiate improvements in the other objectives. The authors of the method, Benayoun et al. (1971) suggest performing a small sensitivity analysis to the Δ-values when the DM is having difficulties in estimating them. To do so, during the calculation phase, say iteration *h*, the analyst should to solve several models with the feasible region at this iteration, $S^{(h)}$, using different values for Δ_j. After seeing the solutions corresponding to different Δ_j, the DM can more easily assess which one best reflects her/his preferences. In other words, the DM will have more knowledge about which value will best improve the unrelaxed objectives.

One important aspect concerning STEM is that it does not assure the compromise solution to be non-dominated, since only the augmented version of the Tchebychev distance has such property (Antunes et al., 2016). Lastly, being a method that makes use of the Payoff matrix, it inherits some of its limitations (see Chapter 7 for details). Namely, the objective functions ranges over the feasible region to be either underestimated or overestimated. The $\alpha_p^{(h)}$ weights are affected if the ranges are not correctly estimated.

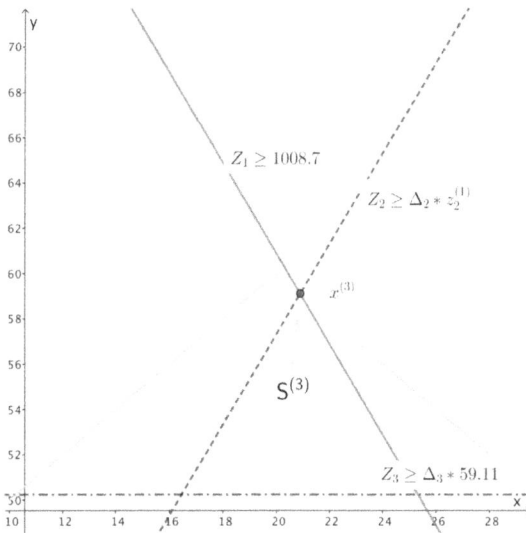

Figure 10.8: Feasible Region in the Decision Space ($S^{(3)}$) and Compromise Solution of STEM Second Iteration, $x^{(3)}$

10.3 Other Methods

As mentioned in the introduction section, STEM is a pioneer method in the fields of interactive multiobjective approaches. It has been superseded by a very large number of other interactive methods. Two decades after it had been published, Shin and Ravindran (1991) reviewed more than a 100 works addressing interactive methods.

Interactive methods may be classified according the way preferences are elicited (Miettinen et al., 2008): trade-off information (e.g. Zionts-Wallenius method developed by Zionts and Wallenius, 1976), reference point approach (e.g. Pareto Race proposed by Korhonen and Laakso, 1986) and classification-based methods (e.g. STEM method). Trade-off is a concept commonly used in decision making but can be tricky to define[7]. Without entering into formal definitions, trade-offs can be viewed as the amount one loses in one objective function when improving the value of another one. Or, as the amount the DM is willing to let go in one objective to improve the value of another one. In trade-off methods, there is always the relation between two values and an opposition of movement. So, most interactive methods based on trade-offs either compute them at each iteration and ask the DM to evaluate their desirability or, are given an efficient solution, and the DM is asked to provide some information concerning the amount (s)he are willing to let go on a given objective (without previously computing them).

In this section, three methods will be described, each one having interesting particularities. The Zionts-Wallenius method explores the objective's space through a weighted sum function and navigates this space with the preferences information assessed through trade-offs. The Pareto Race makes use of visualization techniques in order to help in eliciting the DM's preference that will set out new search directions. Lastly, a quite recently developed method, NAUTILUS, that starts the iterative process at the worst possible solution, so that the DM can feel the solutions improving at stages during the iterative process.

10.3.1 Zionts-Wallenius Method

The Zionts-Wallenius method was proposed by Zionts and Wallenius (1976) and extended in Zionts and Wallenius (1983) to addressed non-linear value functions. It is based on the weighted sum method (studied in Chapter 9). It assumes the DM has a (implicit) value function that guides her/his behavior making the iterative process converge to the efficient extreme point of greatest value. Though, the method does not try to explicitly estimate this value function. It rather guides the search for the compromise solution taking into account the responses provided by the DM in the dialog phase, as it assumes the DM behaves according to this value function. The questions aim at assessing the trade-offs the DM is willing to make with respect to the weights used in the weighted sum function. By comparing two

---

[7] The concept is also discussed in Chapter 3 when addressed the SMART method for multiple criteria problems and in chapter

solutions at a time, the DM indicates her/his preference with respect to one of them. The given response will be translated into a constraint over the weighting space, reducing it. However, since a pairwise comparison of solutions are judgments that are not easy to make, the DM may not be able to make a decision or may be indifferent. In those cases, no constraint is added to the model. At the calculation phase, a new efficient solution is computed. In addition, other efficient extreme points (adjacent to the current solution) are identified and the corresponding trade-offs are presented to the DM. The dialog phase starts once again.

This iterative process stops when the DM prefers the current solution (the one the iteration started with) over the new computed ones. Another stopping criteria comes when the weighting space is reduced to a point so that it is possible to identify one final compromise solution.

Being a weighted sum method, all objective functions have been normalized before starting the iterative process (this issue has been addressed in Chapter 9). For all the details concerning this algorithm refer to Steuer (1986) or Antunes et al. (2016).

10.3.2 Pareto Race

The Pareto Race method was proposed by Weistroffer and Li (1988), not only as an interactive procedure, but also as a visual tool. At the time not enough attention had been paid to the visual display of the large amounts of information present at each iteration. The user interface was one of the main focuses of the authors when developing the Pareto Race.

This method builds upon the reference directional approach proposed by Korhonen and Laakson (1986). In the reference directional approach, a subset of efficient solutions is generated and shown to the DM using a graphic representation. However, this picture was static (one picture per iteration) and did not allow the DM to explore the efficient frontier by deciding, for instance, where to go in the next search (new search direction). The Pareto Race method, using a friendly visual interface, allows the DM to "play" with the search direction and the step value (another of the method's parameters) which enables her/him to immediately see the impact changes will have on the objective function values.

The method starts by asking the DM which values (s)he wish each objective function to achieve. These values do not need to be attainable, since they reflect the DM's aspiration with respect to the objective function values. A non-dominated solution is then computed by minimizing the weighted Tchebychev distance to the aspiration levels. In the Dialog phase, the DM searches the Pareto Front by manipulating some bar-charts showing the objective functions' value length adapting dynamically. At this stage and before committing to an answer about their preferences, the DM is able to perform a sensitivity analysis on the next direction to follow and on which step value to take, evaluating the impact of the changes. At the calculation step, a parametric optimization model is solved to compute a new reference point (Allmendinger et al., 2017).

This method can be of use for problems not having more than 10 objective functions, and constraints and variables in the (few) hundreds (Korhonen, 2005). Larger problems will lead to higher computational times which is very inconvenient in an iterative process where the decision marker is waiting for solutions.

All the details on the Pareto Race method can be found in Korhonen and Wallenius (1988). Regarding other solution visualization methods in MOLP, very interesting works can be found in Korhonen and Wallenius (2008) and, more recently, in Miettinen (2014).

10.3.3 NAUTILUS

The NAUTILUS is a method with a solution philosophy quite different for the majority of interactive methods. It is based on the assumption that the "past experiences affect decision makers' hopes and that people do not react symmetrically to gains and losses" (Miettinen, 2010). This approach is best suited for DMs that prefer to start from the worst possible solution and, progressively, move towards better ones. Therefore, the methodology starts at the nadir point and, at the calculation phase, determines a new solution that dominates the previous one. Moreover, the method is *trade-off-free* and so the DM does not need to make hard decisions as some of the previous methods ask for.

The first question the DM is asked is how many iterations are (s)he willing to carry out, setting in this manner the stopping criteria[8]. At the dialog phase the DM may either define the direction of improvement of the objectives or (s)he choose one solution among some alternatives provided by the facilitator. At the calculation phase, an augmented weighted Tchebychev distance model is solved and a solution that dominated all previous ones is calculated. The consequence of being a trade-off-free method is that the range of attainable values for the objective functions shrinks as the iterations evolve. To provide supporting information for the DM at the dialog phase, lower and upper bounds on these reachable values are computed when possible. The ε-constraint method is the approach applied to calculate these bounds.

Since it was first published by Miettinen et al. (2010), this method has grown into a family of NAUTILUS methods. Miettienen and Ruiz (2016) describe with detail this method framework and present four different variants.

10.4 Final Remarks

This chapter addresses methods where the preference elicitation component is based on an interactive tool. One crucial aspect for using interactive methods is the time and willingness of the DM must have to be part of the iterative process, since (s)he bring fundamental information for the process to evolve. When the process finishes and a compromise solution has been found, the DM has had the

---

[8] The DM is allowed to revise this number during the iterative process.

possibility to learn deeply about the problem (s)he faces and is now more capable to justify why the final solution is the best one. This feeling of ownership with respect to the compromise solution is something very hard to develop in the DM if *a priori* or *a posterior* approaches are used[9].

Over the years, a very large number of methods have been proposed. None of them can be called superior to any other (Miettinen and Hakanen, 2017). In fact, the "best" method depends on several factors: how comfortable the DM feels with respect with the preference information (s)he needs and are capable of providing; the type of compromise solution one can obtain (e.g. Pareto optimal solution, weakly efficient solution); the mathematical assumptions and properties of the problems, among many others. Given this problem's dependency and the number of methods that could have been presented in this chapter, the choice was to address with detail one of the first methods published in this class of models (STEM method). This method has a simple calculation phase which makes it a good pedagogical tool.

The reader should refer to comprehensive books from Ehrgott (2005), Miettinen (1998) and Greco et al. (2016), if interested in other methods and/or their mathematical fundamentals.

10.5 Exercises

10.5.1 Solved Exercises

Consider the MOPL problem with three objective functions presented below and apply the STEM method to determine a compromise solution. Notice objective Z_2 is to be minimized. The STEM procedure presented above assumes that all objectives are to be maximized. Therefore, to make the method application more straightforward, one should change the second objective optimization direction:

$$\text{Max } Z_2{}' = x - y.$$

From this point on, when questioning the DM regarding the second objective function, the facilitator should carefully make the question since the DM will, for sure, think of it as a minimization function and not as a maximization one as being solved.

$$\text{Max } Z_1 = x + 2y$$
$$\text{Max } Z_2{}' = x - y$$
$$\text{Max } Z_3 = 3x - y$$
$$\text{s.t. } x + 3y \geq 6$$
$$2x - y \leq 5$$
$$-x + y \leq 3$$
$$x + y \leq 7$$
$$x, y \geq 0$$

---

[9] These methods have been presented in Chapters 8 and 9.

Firstly, one should compute the payoff table (Table 10.3). The ideal and nadir solutions are, respectively, $Z^* = (12, 2, 9)$ and $N^* = (5, -3, 1)$. The corresponding β values are

Table 10.3: Payoff Table

	Z_1	Z_2'	Z_3
$x_1^* = (2, 5)$	12	-3	1
$x_2^* = (3, 1)$	5	2	8
$x_3^* = (4, 3)$	10	1	9

$$\beta_1 = 0.2609, \ \beta_2 = 1.7678 \text{ and } \beta_3 = 0.2811.$$

For the first iteration $I^* = \varnothing$ and $h = 1$. The corresponding α-values are

$$\alpha_1^{(1)} = 0.113, \ \alpha_2^{(1)} = 0.7653 \text{ and } \alpha_3^{(1)} = 0.1217.$$

The model to be solve at this moment is

$$\text{Min } Z = \upsilon$$
$$\text{s.t.} \ \ x + 3y \geq 6$$
$$2x - y \leq 5$$
$$-x + y \leq 3$$
$$x + y \leq 7$$
$$\upsilon \geq \alpha_1^{(1)} [12 - (x + 2y)]$$
$$\upsilon \geq \alpha_2^{(1)} [2 - (x - y)]$$
$$\upsilon \geq \alpha_3^{(1)} (9 - (3x - y)$$
$$x, y, \upsilon \geq 0$$

The first compromise solution is $(x, y) = \left(\dfrac{11}{3}, \dfrac{7}{3}\right)$ to with corresponds $Z_1 = 8.333$, $Z_2' = 1.333$ and $Z_3 = 8.667$. The method progresses with the dialog phase as presented in Example 10.3.

10.5.2 Proposed Exercises

1. Consider the MOLP problem

$$\text{Max } Z_1 = y$$
$$\text{Max } Z_2 = x - 2y$$
$$\text{s.t.} \ \ \ \ 5x - 2y \leq 40$$
$$3x + 5y \leq 55$$
$$-x + y \leq 3$$
$$x + y \geq 3$$
$$x, y \geq 0$$

(a) Solve independently each objective function presenting the optimal solution (or optimal solutions) and the corresponding optimal value. Graph the problem on the objectives space and determine the set of non-dominated solutions.

(b) Compute the payoff table.

(c) Determine the ideal solution and the nadir point.

(d) Compute the β values and the corresponding α.

(e) Formulate the model corresponding to STEM's first iteration and, using an optimization software (e.g. GAMS), compute the corresponding compromise solution.

(f) Graph the previous model in the objectives space and verify if the previously computed compromise solution is an efficient solution.

(g) Assume the DM is willing to relax 10% of the first objective function optimal values. Formulate the model corresponding to STEM's second iteration.

(h) Graph, in the objectives space, the previous model and determine the corresponding compromise solution.

(i) Now, assume the DM is willing to relax 10% of the second objective function optimal value. Formulate the model corresponding to STEM's second iteration.

(j) Graph, in the objectives space, the previous model and determine the corresponding compromise solution.

2. Consider the MOLP problem

$$\text{Max } Z = (2x_1 + x_2; x_1 - x_2)$$
$$\begin{aligned}
\text{s.t.} \quad & x_1 + x_2 \geq 2 \\
& -2x_1 + x_2 \geq 0 \\
& -x_1 + x_2 \leq 4 \\
& x_1, x_2 \geq 0
\end{aligned}$$

(a) Solve independently each objective function presenting the optimal solution (or optimal solutions) and the corresponding optimal value. Graph the problem on the objectives space and determine the set of non-dominated solutions.

(b) Compute the payoff table.

(c) Determine the ideal solution and the nadir point.

(d) Compute the β values and the corresponding α.

(e) Formulate the model corresponding to STEM's first iteration and, using an optimization software (e.g. GAMS), compute the corresponding compromise solution.

(f) Graph the previous model in the objectives space and verify the previously computed compromise solution is an efficient solution.

(g) Assume the DM is willing to relax 10% of the first objective function optimal value. Formulate the model corresponding to STEM's second iteration.

(h) Graph, in the objectives space, the previous model and determine the corresponding compromise solution.

(i) Now, assume the DM is willing to relax 10% of the second objective function optimal value. Formulate the model corresponding to STEM's second iteration.

(j) Graph, in the objectives space, the previous model and determine the corresponding compromise solution.

3. Consider the MOLP problem

$$\text{Max } Z_1 = x - y$$
$$\text{Max } Z_2 = -2x + y$$
$$\text{s.t.} \quad x \geq 3$$
$$x + y \leq 12$$
$$x + y \geq 4$$
$$x - y \leq 6$$
$$x, y \geq 0$$

(a) Solve independently each objective function presenting the optimal solution (or optimal solutions) and the corresponding optimal value. Graph the problem on the objectives space and determine the set of non-dominated solutions.

(b) Compute the payoff table.

(c) Determine the ideal solution and the nadir point.

(d) Compute the β values and the corresponding α.

(e) Formulate the model corresponding to STEM's first iteration and, using an optimization software (e.g. GAMS), compute the corresponding compromise solution.

(f) Graph the previous model in the objectives space and verify the previously computed compromise solution is an efficient solution.

(g) Assume the DM is willing to relax 15% of the first objective function optimal value. Formulate the model corresponding to STEM's second iteration.

(h) Graph, in the objectives space, the previous model and determine the corresponding compromise solution.

(i) Now, assume the DM is willing to relax 15% of the second objective function optimal value. Formulate the model corresponding to STEM's second iteration.

(j) Graph, in the objectives space, the previous model and determine the corresponding compromise solution.

4. Consider the MOLP problem

$$\text{Max } Z_1 = 2x_1 + x_2$$
$$\text{Min } Z_2 = -x_1 + x_2$$
$$\text{s.t.} \quad x_1 + x_2 \geq 2$$
$$-2x_1 + x_2 \geq 0$$
$$-x_1 + x_2 \leq 4$$
$$x_1, x_2 \geq 0$$

(a) Solve independently each objective function presenting the optimal solution (or optimal solutions) and the corresponding optimal value. Graph the problem on the objectives space and determine the set of non-dominated solutions.

(b) Compute the payoff table.

(c) Determine the ideal solution and the nadir point.

(d) Compute the β values and the corresponding α.

(e) Formulate the model corresponding to STEM's first iteration and, using an optimization software (e.g. GAMS), compute the corresponding compromise solution.

(f) Graph the previous model in the objectives space and verify the previously computed compromise solution is an efficient solution.

(g) Assume the DM is willing to relax 10% of the first objective function optimal value. Formulate the model corresponding to STEM's second iteration.

(h) Graph, in the objectives space, the previous model and determine the corresponding compromise solution.

(i) The DM wants to revise the judgement made for the first objective. Now, the DM wishes to relax, instead, 10% of the second objective function optimal value. Formulate the model corresponding to this new STEM's second iteration.

(j) Graph, in the objectives space, the previous model and determine the corresponding compromise solution.

5. Consider the MOLP problem

$$\text{Max } Z_1 = -x + 3y$$
$$\text{Max } Z_2 = 3x + 3y$$
$$\text{Max } Z_3 = x + 2y$$
$$\text{s.t.} \quad y \leq 4$$
$$x + 2y \leq 10$$
$$2x + y \leq 10$$
$$x, y \geq 0$$

(a) Solve independently each objective function presenting the optimal solution (or optimal solutions) and the corresponding optimal value. Graph the problem on the objectives space and determine the set of non-dominated solutions.

(b) Compute the payoff table.

(c) Determine the ideal solution and the nadir point.

(d) Compute the β values and the corresponding α.

(e) Formulate the model corresponding to STEM's first iteration and, using an optimization software (e.g. GAMS), compute the corresponding compromise solution.

(f) Assume the DM is willing to relax 10% of the second objective function optimal value. Formulate the model corresponding to STEM's second iteration.

(g) Now, assume the DM is willing to relax 5% of the first objective function optimal value. Formulate the model corresponding to STEM's third iteration.

(h) Proceed as previously and using STEM algorithm, determine a solution of compromise testing different values for the Δ_k.

6. Consider the MOLP problem

$$\text{Max } Z_1 = -x + 3y$$
$$\text{Max } Z_2 = 3x + 3y$$
$$\text{Max } Z_3 = x + 2y$$
$$\text{s.t.} \qquad y \leq 4$$
$$x + 2y \leq 10$$
$$2x + y \leq 10$$
$$x, y \geq 0$$

Present the formulation of the model one needs to solve to find the compromise solution given by the first iteration of the STEM algorithm. Show all the necessary calculations to get to this formulation.

7. Consider the MOLP problem

$$\text{Max } Z_1 = x_1 + x_2$$
$$\text{Max } Z_2 = x_1 - x_2$$
$$\text{Max } Z_3 = -x_1 + x_2$$
$$\text{s.t.} \qquad -x_1 + x_2 \leq 4$$
$$x_2 \leq 6$$
$$2x_1 - x_2 \leq 8$$
$$x_1, x_2 \geq 0$$

(a) Find graphically the optimal solution of each objective function.

(b) Compute the payoff table.

(c) Determine the ideal solution and the nadir point.

(d) Aiming to apply STEM method, the β-values were computed

$$\beta_1 = 0.49, \beta_2 = \beta_3 = 1.414.$$

The compromise solution found at STEM first iteration was $(x_1, x_2) = (6, 6)$. The DM has decided to relax Z_1 at most 2 units. Propose the STEM model formulation for the next iteration.

8. Consider the MOLP problem

$$\text{Max } Z_1 = x_1 + x_2$$
$$\text{Max } Z_2 = -2x_1 + x_2$$
$$\text{Max } Z_3 = 2x_1 + x_2$$
$$\text{s.t.} \qquad x_1 + x_2 \leq 12$$
$$x_1 \geq 3$$
$$x_1 - x_2 \leq 0$$
$$x_1, x_2 \geq 0$$

(a) Find graphically the optimal solution of each objective function.
(b) Compute the payoff table.
(c) Determine the ideal solution and the nadir point.
(d) Aiming to apply STEM method, the β-values were computed

$$\beta_1 = 0.353, \ \beta_2 = 0.391 \text{ and } \beta_3 = 0.280.$$

The compromise solution found at STEM first iteration was $(x_1, x_2) = (6.35, 6.65)$. The DM has decided to relax Z_3 at most 3 units. Propose the STEM model formulation for the next iteration.

9. Consider the MOLP problem

$$\text{Max } Z_1 = x_1 + 2x_2$$
$$\text{Max } Z_2 = -2x_1 + x_2$$
$$\text{Max } Z_3 = 2x_1 + x_2$$
$$\text{s.t.} \qquad x_1 + x_1 \leq 16$$
$$-x_1 + x_1 \leq 0$$
$$x_1 \geq 2$$
$$x_1 - 2x_2 \leq 4$$
$$x_1, x_2 \geq 0$$

(a) Represent graphically the decision space and determine the solutions that might be of interest to the Decision Maker.
(b) Determine the ideal solution and the nadir point.
(c) Find the compromise solution when none of the objective functions have been relaxed.
(d) Propose the STEM model formulation for the next iteration, knowing the DM is willing to relax Z_3 about 10% of the value found in the previous exercise.
(e) Using an optimization software, compute the compromise solution corresponding to the previous formulated model.

10. Aunt Leonarda recently bought 22 kilograms of cregola, a very rare substance used in perfumery. She wants to use it to produce her two most famous perfumes, named as X and Y, for reasons of confidentiality. Aunt Leonarda wonders: "Each kiloliter of X gives me a profit of one thousand pilins, while with the perfume Y I manage to collect profits of 3 thousand

pilins per kiloliter. But, each kiloliter of Y consumes two kilos of cregola, and with X the consumption is one to one...."

"Don't forget that to fulfil the production plan, at least three kiloliters of X and two of Y should be produced ,... and that our distributor can only assure selling eight kiloliters of Y",- added Blimunda, Aunt's secretary. – "And there is something else ,... the total of X produced has to exceed the total production of Y by six kiloliters" - Blimunda reminded her.

"What a mess,... I want to minimize the consumption of cregola with this production, but I also want to make the highest profit." Admitted the friendly Aunt Leonarda.

Blimunda wrote something on paper and proudly presented the following formulation to Aunt Leonarda.

$$\text{Max } Z(-x_1 - x_2; x_1 + 3x_2)$$
$$\text{s.t.} \qquad x_1 \geq 3$$
$$x_2 \geq 2$$
$$x_2 \leq 8$$
$$x_1 - x_2 \leq 6$$
$$x_1 + 2x_2 \leq 21$$

Using an optimization software (e.g. GAMS) apply the STEM algorithm and determine a solution of compromise testing different values for Δ_k.

11. Jaime is preparing his study time for next week. He will have two tests (Algebra and Biochemistry). For each study hour of Algebra his girlfriend will offer him 5 dl of strawberry ice cream. But she already threatened him that she would reduce 2 dl of ice cream for every hour he studied Biochemistry. Note that his girlfriend is a Math student... Every hour of Biochemistry study will give him 4 units of satisfaction. But an hour of Algebra study will provide him with a unit of negative satisfaction ...

   Jaime does not want to study more than 6 hours of Algebra, but the amount of Biochemistry study hours should not exceed in 3 hours the Algebra study period. He would like to maximize the amount of ice cream he will receive, but also the total satisfaction with the study.

   This problem can be formulated as

   $$\text{Max } Z_1 = 5x - 2y$$
   $$\text{Max } Z_2 = -x + 4y$$
   $$\text{s.t.} \qquad -x + y \leq 3$$
   $$x \leq 6$$
   $$x, y \geq 0$$

   (a) Describe the meaning of the variables, objective functions and constraints.
   (b) Graph the feasible region in the decision space and in the objectives space, marking the set of efficient solutions and the set of non-dominated solutions.

(c) Build the payoff table and determine the ideal and nadir points.
(d) Formulate the problem corresponding to STEM first iteration. Determine the compromise solution. Represent the solution on the feasible region.
(e) Assume Jaime does not mind letting go of 1 dl of the optimal value for the ice-cream quantity. In accordance with the STEM algorithm, formulate the problem that models Jaime's preference.
(f) Represent the new constraint on the feasible region and determine the new compromise solution
(g) Using an optimization software (e.g. GAMS) apply the STEM algorithm and determine a compromise solution testing different values for Δ_k.

12. In a factory there are two types of machines used in the production lines of four products. The available capacities of each type of machine (in hours / week) and the number of processing hours/week required at each machine, per unit produced product, are presented in the table:

Machine	P_1	P_2	P_3	P_4	Availability
A	2	1	4	3	60
B	3	4	1	2	60

The company management wants to maximize profit, quality of P_1, P_2, P_3 and P_4 supply and the degree of employee satisfaction. Recent studies have made it possible to establish the unit contribution of each product to each objective:

	P_1	P_2	P_3	P_4
Profit	3	1	2	1
Quality	1	−1	2	4
Satisfaction	−1	5	1	2

Using an optimization software (e.g. GAMS) apply the STEM algorithm and determine a solution of compromise testing different values for Δ_k.

References

Allmendinger, R., M. Ehrgott, X. Gandibleux, M.J. Geiger, K. Klamroth and M. Luque (2017). Navigation in multiobjective optimization methods. *Journal of Multi-Criteria Decision Analysis*, 24(1-2), 57–70.

Antunes, C.H., M.J. Alves and J. Climaco (2016). *Multiobjective Linear and Integer Programming*. Springer, New York, NY.

Benayoun, R., J. De Montgolfier, J. Tergny and O. Laritchev (1971). Linear programming with multiple objective functions: Step method (STEM). *Mathematical Programming*, 1(1), 366–375.

Ehrgott, M. (2005). *Multicriteria Optimization* (Vol. 491). Springer Science & Business Media.

Greco, S., J. Figueira and M. Ehrgott (2016). *Multiple Criteria Decision Analysis* (Vol. 37). New York: Springer.

Haimes, Y.Y., L.S. Lasdon and D.A. Wismer (1971). On a bicriterion formulation of the problems of integrated system identification and system optimization. *IEEE Trans. Syst. Man. Cybern.*, 1, 296–297.

Korhonen, P. (2005). Interactive methods. *In:* Figueira, J., S. Greco and M. Ehrgott (Eds.), Multiple Criteria Decision Analysis (pp. 641–665). State of the Art Surveys.

Korhonen, P.J. and J. Laakso (1986). A visual interactive method for solving the multiple criteria problem. *European Journal of Operational Research*, 24(2), 277–287.

Korhonen, P. and J. Wallenius (1988). A Pareto race. *Naval Research Logistics (NRL)*, 35(6), 615–623.

Korhonen, P. and J. Wallenius (2008). Visualization in the multiple objective decision-making framework. *In:* Multiobjective Optimization (pp. 195–212). Springer, Berlin, Heidelberg.

Miettinen, K. (1998). *Nonlinear Multiobjective Optimization* (Vol. 12). Springer Science & Business Media.

Miettinen, K. (2014). Survey of methods to visualize alternatives in multiple criteria decision making problems. *OR Spectrum*, 36(1), 3–37.

Miettinen, K. and J. Hakanen (2017). Why use interactive multi-objective optimization in chemical process design? *In:* G.P. Rangaiah (Ed.), Multi-Objective Optimization: Techniques and Applications in Chemical Engineering (pp. 157–197). Advances in Process Systems Engineering, 5. World Scientific.

Miettinen, K. and F. Ruiz (2016). NAUTILUS framework: Towards trade-off-free interaction in multiobjective optimization. *Journal of Business Economics*, 86(1–2), 5–21.

Miettinen, K., P. Eskelinen, F. Ruiz and M. Luque (2010). NAUTILUS method: An interactive technique in multiobjective optimization based on the nadir point. *European Journal of Operational Research*, 206(2), 426–434.

Miettinen, K., F. Ruiz and A.P. Wierzbicki (2008). Introduction to multiobjective optimization: Interactive approaches. *In:* Multiobjective Optimization (pp. 27–57). Springer, Berlin, Heidelberg.

Shin, W.S. and A. Ravindran (1991). Interactive multiple objective optimization: Survey I - Continuous case. *Computers & Operations Research*, 18(1), 97–114.

Steuer, R.E. (1986). *Multiple Criteria Optimization. Theory, Computation and Applications.* John Wiley and Sons, Inc.

Tamiz, M. and D.F. Jones (1997). Interactive frameworks for investigation of goal programming models: Theory and practice. *Journal of Multi-Criteria Decision Analysis*, 6(1), 52–60.

Winston, W.L. (2002). *Introduction to Mathematical Programming: Applications and Algorithms*, Duxbury.

Zionts, S. and J. Wallenius (1976). An interactive programming method for solving the multiple criteria problem. *Management Science*, 22(6), 652–663.

Zionts, S. and J. Wallenius. (1983). An interactive multiple objective linear programming method for a class of underlying nonlinear utility functions. *Management Science*, 29(5), 519–529.

Appendix A

Models in GAMS

*GAMS code for Example 10.2

positive variables x, y, v;
variable z;

equations
eq1, eq2, eq3,eq4,
eqz1, eqz2,
fo;

eq1.. $x + y = g = 20$;
eq2.. $2*x - y = l = 40$;
eq3.. $-x + y = l = 40$;
eq4.. $x + y = l = 80$;
eqz1.. $0.678*(1200 - (20*x + 10*y)) = l = v$;
eqz2.. $0.322*(1600 - (-80*x + 40*y)) = l = v$;
fo.. $z = e = v$;

Model Exa102 /all/;
Option LP = CPLEX;
solve Exa102 using LP minimizing z;

```
* GAMS code for Example 10.3

scalars
a1  "alfa1" /0.2222/
a2  "alfa2" /0.6667/
a3  "alfa3" /0.1111/;

variable z;
positive variables x, y, v;

equations
eq1, eq2, eq3,
eqz1, eqz2, eqz3,
fo;

eq1.. x + y = g = 20;
eq2.. 2*x – y = l = 40;
eq3.. –x + y = l = 40;
eq4.. x + y = l = 80;
eqz1.. a1*(1200 – (20*x + 10*y)) = l = v;
eqz2.. a2*(1600 – (–80*x + 40*y)) = l = v;
eqz3.. a3*(60 – y) = l = v;
fo..  z = e = v;

Model Exa103 /all/;
Option LP=CPLEX;
solve Exa103 using LP minimizing z;
```

CHAPTER

11

Goal Programming

11.1 Introduction

Goal programming (GP) is a mathematical modeling approach for multiobjective optimization which involves the use of aspiration levels (Wierzbicki 1982) instead of objectives as in the previous chapters. In GP, the Decision Maker (DM) sets some targets (s)he want to achieve, and the main objective will be reaching them as far as possible (Romero and Rehman 1984).

This technique was first applied by Charnes et al. (1955) when addressing the problem of defining a compensation plan for executives in an industrial setting. The authors proposed it as an extension of Linear Programming which was able to solve problems that more common methods at the time (e.g. least squares minimization) were finding difficult or even impossible to apply. However, the term "Goal Programming" only appeared in Charnes and Cooper's book (1961) and, according to Schniederjans (2012), had a "humble beginning" since it was not even included as a term in the book's index. Nowadays, GP continues to be very popular in engineering, management science, and economics and social science. In a recent review, Colapinto et al. (2017) collected more than 300 papers published mostly after 2000. The authors concluded that the number of publications where GP is applied keeps growing and that this technique is valued by the academy more than ever.

This chapter is an introduction to Goal Programming modelling. The aim is to set the baseline so that an interested reader can deepen her/his knowledge in what concerns the mathematical fundamentals and the solution procedures in more advanced and dedicated books (e.g. Jones and Tamiz 2010, Schniederjans 1995).

The organization of this chapter is as follows. Firstly, the main concepts in GP such as goals and utopian solutions are defined (Section 11.2). Then, the two classical GP models are presented with detail: the Preemptive model (Section 11.3.1) and the Archimedean model (11.3.2). Next, in Section 11.3.3, two other GP models will be presented as examples of how the two classical models are in fact the basis for most other GP models. Some final remarks are drawn, highlighting some limitations of this modelling approach (Section 11.4). This chapter ends with a selection of proposed exercises covering the addressed topics.

11.2 Goals and the Utopian Set

The setting and attainment of goals is a central aspect of decision making. Although a DM may have different types of goals, most of them will fall into one of four different types: greater than or equal to; less than or equal to; between two values, and equality.

These goals may be mathematically expressed as

- $Z_1 \geq t_1$
- $Z_2 \leq t_2$
- $t'_3 \leq Z_3 \leq t_3$
- $Z_4 = t_4$

where t_i are the values set by the DM as the smallest to be attained, the largest to be attained, exactly the value to be attained and between which (s)he would like to attain. These values are called **targets**. In short, the main idea behind Goal Programming is to find, among all feasible solutions, the one that is "closer" to the targets set by the DM.

Let's take the example that have been accompanying the previous chapters to see how GP differs from the (more traditional) multiobjective optimization model. A fourth "objective" will be added so that all four types of goals can be exemplified.

Example 11.1

The DM aims at determining the best solution having four different perspectives over the problem. The solution (s)he wish to find should reach four different targets:

- $Z_1 = 20x + 10y$ should be at least 700
- $Z_2 = -80x + 40y$ should not be higher than 900
- $Z_3 = y$ should be between 35 and 50
- $Z_4 = 4x - y$ should be 45.

These wishes can be translated into goals as

- $20x + 10y \geq 700$
- $-80x + 40y \leq 900$
- $35 \leq y \leq 50$
- $4x - y = 45.$

The intersection of the first three goals, over the decision space[1], is depicted in Fig. 11.1. Adding the fourth goal, the intersection will now be the segment of line $4x - y = 45$ that falls within the grey area in Fig. 11.1 (the black segment in Fig. 11.2).

---

[1] The decision space is the space defined by the decision variables.

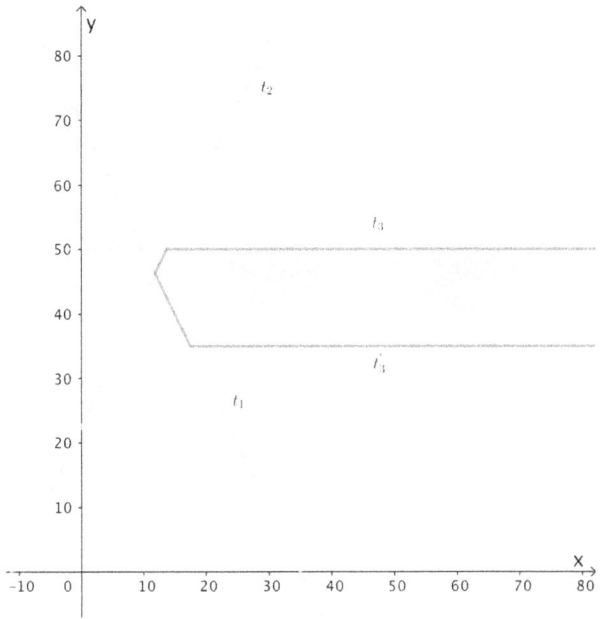

Figure 11.1: Intersection of the First Three Goals in the Decision Space

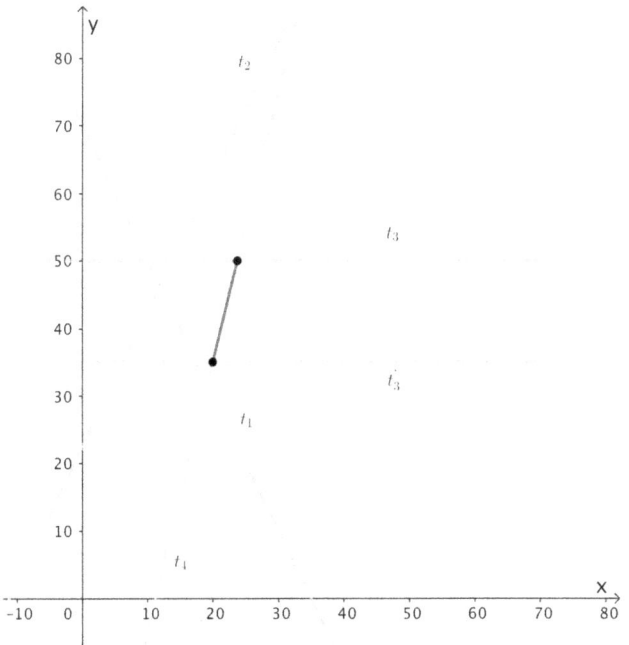

Figure 11.2: Intersection of the Four Goals in the Decision Space (Utopian Set)

In goal programming, the space defined by all goals is known as the **utopian set**. This set contains all the points in the decision space (R^2 in this example) simultaneously satisfying all the targets.

Assume the DM seeks the solution in a feasible region defined by a group of constraints (set S) that reaches all targets set for the goals. The feasible region (set S) in the decision space is shown in Fig. 11.3 alongside the constraints that define it.

$x + y \geq 20$

$2x - y \leq 40$

$-x + y \leq 80$ (S)

$x + y \leq 80$

$x, y \geq 0$

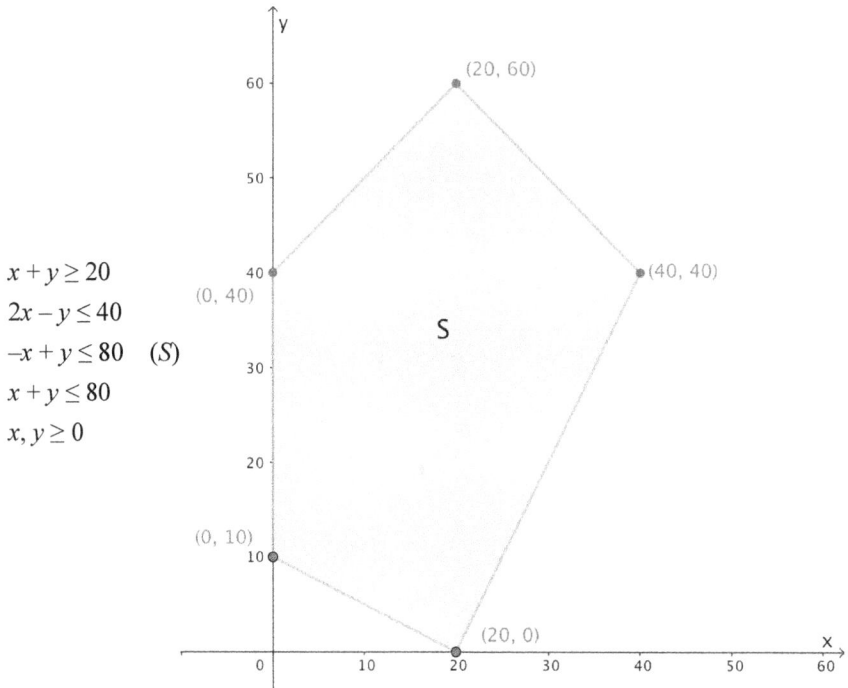

Figure 11.3: Feasible Region in the Decision Space (set S)

The goal programming model formulation is given below. Figure 11.4 shows the utopian space projected onto the feasible region. One can see that all utopian solutions are also feasible solutions. Therefore, for this DM, any utopian solution is as good a solution as any other, since they all meet all the targets. (S)he can choose any point of the utopian set as the problem's (final) solution.

The utopian set in this example has two characteristics. First, it is a non-empty set, meaning that all the goals intersected with each other. Second, it intersects the feasible region which indicates that there are solutions that are both utopian solutions and feasible solutions. However, this is not always the case. In the two subsections below, one will firstly address cases where the feasible region and the utopian set do not intersect. Then the case in which the utopian set is empty will be studied.

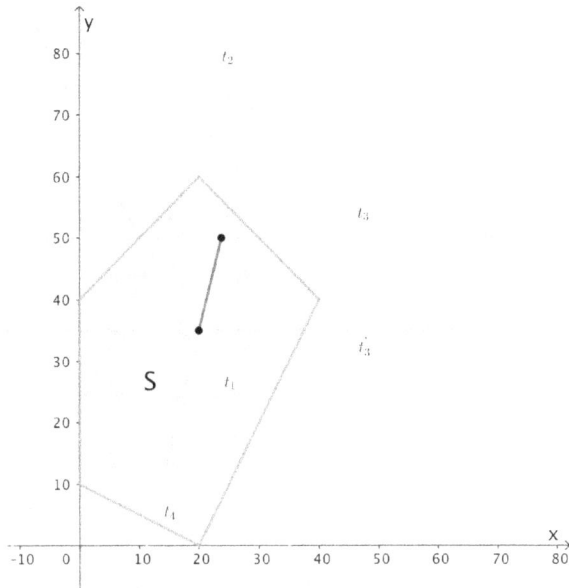

Figure 11.4: Projection of the Utopian Space over the Feasible Region (*S*)

Goal $(20x + 10y \geq 700)$
Goal $(-80x + 40y \leq 900)$
Goal $(35 \leq y \leq 50)$
Goal $(4x - y = 45)$
s.t. $x + y \geq 20$
       $2x - y \leq 40$
       $-x + y \leq 80$
       $x + y \leq 80$
       $x, y \geq 0$

11.2.1 What if the Utopian Set Does Not Intersect the Feasible Region?

The targets may be such that none of the solutions in the utopian set are feasible solutions. This is the case when the feasible and the utopian sets do not intersect. In such situation, one should seek for the feasible solution that is closer to the utopian set.

Example 11.2

Suppose the target for $Z_3 = y$ is between 65 and 80, while the targets for the other goals remain unchanged. Such a change leads to a utopian set that does not intersect the feasible region (Fig. 11.5).

For the problem to be solved, the DM needs to provide some more information. This information concerns the importance that the different goals have to the DM. The more important a goal is the highest priority it must be given. Higher priority goals should be met as much as possible and only the least priority goals may have some deviation from the target. Figure 11.5 shows two candidates for the problem solution: $(x, y) = (20, 60)$ and $(x, y) = (25, 55)$. Both are feasible solutions that are close to the utopian set. So, which one is the best one? The answer is "it depends". Different priorities will make one "better" than the other. Note, that if DM gives priority to goal 4 over goal 3, then the solution should be $(x, y) = (25, 55)$ which has zero deviation from goal 4 and 10 units of deviation from goal 3. On the contrary, if a higher priority is given to goal 3 over goal 4, then the solution

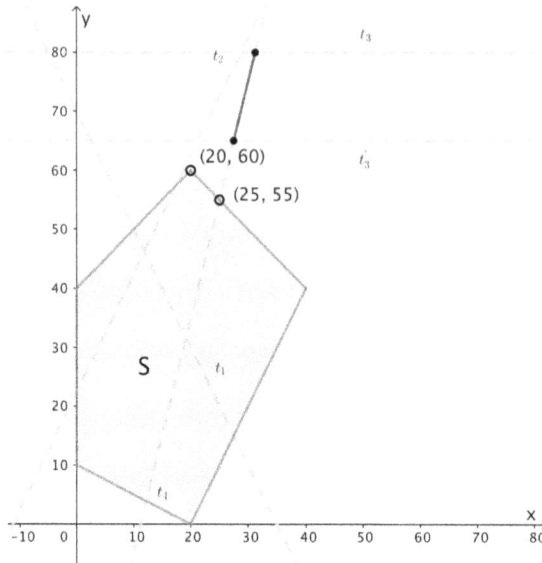

Figure 11.5: Possible Optimal Solution for Example 11.2

should be $(x, y) = (20, 60)$ since it has a smaller deviation to the target (5 units instead of 10).

To sum up, when the utopian and the feasible sets do not intersect, one needs to take the DM's preferences concerning the goals, they will influence the final solution.

11.2.2   What if the Utopian Set is an Empty Set?

The targets set by the DM may be such that the utopian set itself is empty. The GP problem will still be possible to solve since one is seeking a feasible solution that best "achieves" the goals even if some of the targets are not met. Again, one needs to know how important it is that the DM evaluates each goal.

Example 11.3

Take the GP problem below where the DM has three goals with $(x, y) \in S$.

- Goal 1: $Z_1 = 20x + 10y$ should not be greater than 650,
- Goal 2: $Z_2 = -80x + 40y$ should be smaller or equal to 400, and
- Goal 3: $Z_3 = y$ shows be between 45 and 65.

Figure 11.6 shows the three regions defined by the three goals. One can see that the goals intersect each other two by two, but not the three simultaneously.

Let's assume the DM considers goal 1 to have priority over goal 2, which in turn has priority over goal 3,

$$Z_1 \gg Z_2 \gg Z_3$$

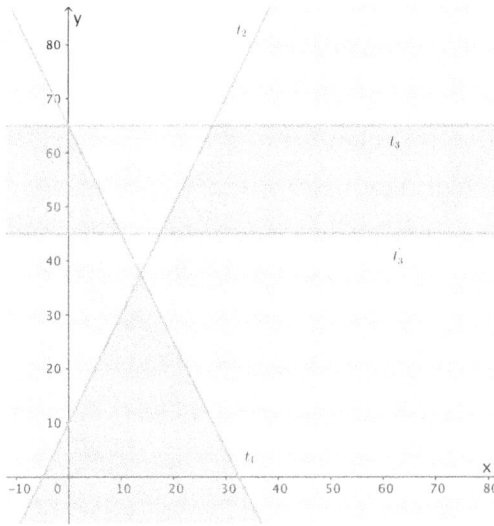

Figure 11.6: Regions Defined by Each Goal

where ">>" reads as "has priority over". Taking the feasible region (set S), one should start by taking the highest priority goal and investigate if it intersects S. Figure 11.7 shows that for goal 1 ($20x + 10y \leq 650$), this intersection is non-empty. In this step, the feasible region is "reduced" to the feasible solutions that meet goal 1. Now, one can proceed in a similar way taking the second priority goal. Figure 11.8 shows all the feasible solutions that simultaneously meet goals 1 and 2. Would this set be different if instead of $Z_1 >> Z_2$, one had $Z_2 >> Z_1$?

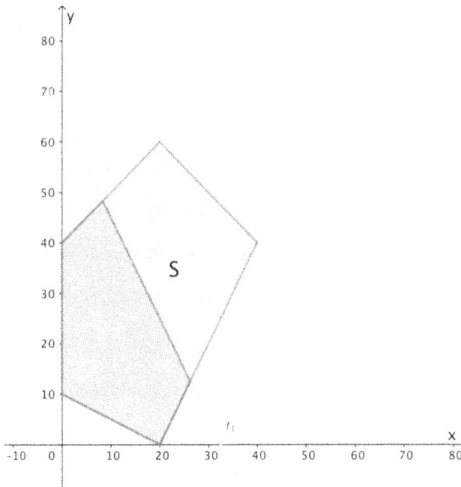

Figure 11.7: Intersection between the Halfplane Defined by Goal 1 and the Feasible Region (set S)

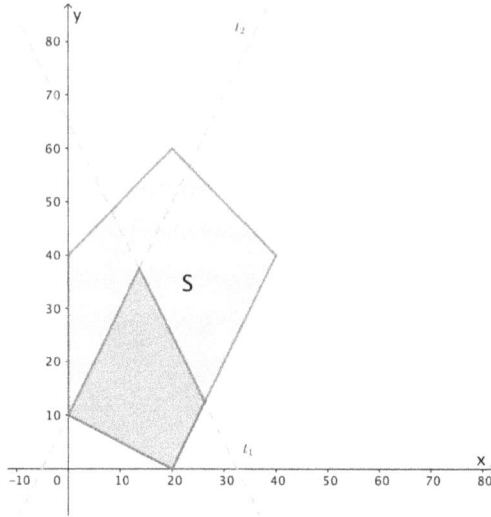

Figure 11.8: Intersection between Goals 1 and 2 and the Feasible Region (set S)

Taking the last goal, $y \geq 63$, one knows it will not intersect the feasible region nor will it have any solution in common with the other two goals. Therefore, one should seek a solution, in the dark grey polyhedron on Fig. 11.8, that is closer to goal 3.

Figure 11.9 shows the optimal solution is $(x, y) = (13.75, 37.5)$. This solution meets goals 1 and 2 and has some deviation from goal 3, which conforms with the priorities set by the DM.

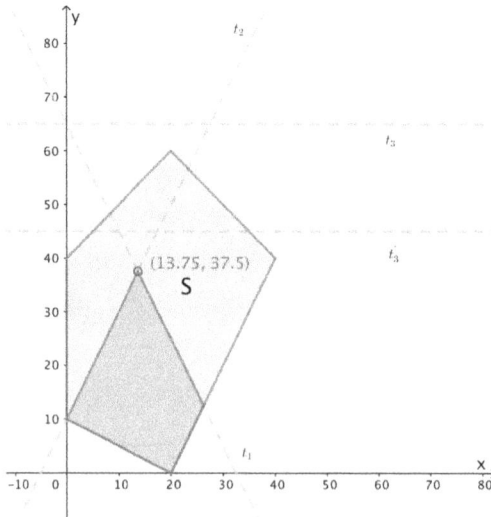

Figure 11.9: Optimal solution of Example 11.3

11.3 Goal Programming Models

A goal programming model seeks to find the solution that minimizes the total deviation from the targets. For each goal, two non-negative deviation variables are added: one will represent the *below* deviation from the target (modelling the cases when the target level is not reached), while the other will represent the *above* deviation from the target (when the target level is surpassed). Amongst these deviation variables, only those presenting *undesirable* deviations should be minimized.

Let's take the four generic goals

- $Z_1 \geq t_1$,
- $Z_2 \leq t_2$,
- $t'_3 \leq Z_3 \leq t_3$,
- $Z_4 = t_4$.

As mentioned, for each one of them two non-negative deviation variables are added:

- $Z_1 + d_1^- - d_1^+ = t_1$,
- $Z_2 + d_2^- - d_2^+ = t_2$,
- $Z_3 + d_3'^- - d_3'^+ = t'_3$,
- $Z_3 + d_3^- = d_3^+ = t_3$,
- $Z_4 + d_4^- - d_4^+ = t_4$.

By adding these variables, one has transformed four goals into five constraints (notice that goal 3 is, in fact, a *double* goal). These five constraints differ from the traditional constraints that define the feasible region as they do not restrict this region. In fact, deviations (*good* or *bad*) from the targets are allowed. This kind of constraint is often named as **soft** constraints in opposition to the **hard** constraints defining the feasible region. Soft constraints act on the objective function since undesirable deviation variables should be minimized.

Formally, a GP model may have the following mathematical formulation:

$$Min\ G\ (d_1^-;d_2^+;d_3'^- + d_3^+;d_4^- + d_4^+)$$

$$s.t.\ Z_1(x) + d_1^- - d_1^+ = t_1$$
$$Z_2(x) + d_2^- - d_2^+ = t_2$$
$$Z_3(x) + d_3'^- - d_3'^+ = t'_3$$
$$Z_3(x) + d_3^- - d_3^+ = t_3$$
$$Z_4(x) + d_4^- - d_4^+ = t_4$$
$$d_1^-, d_1^+, d_2^-, d_2^+, d_3'^-, d_3'^+, d_4^-, d_4^+ \geq 0$$
$$x \in S$$

Note that, for each type of goal, different deviation variables are being minimized. Actually, only the *undesirable* deviations should be considered. Function *G* brings together these unwanted deviation variables. Very often in scientific literature, this function is named as the **achievement function** (e.g. Romero 1991, Jones and Tamiz 2010) to avoid misinterpretations with the traditional objective function. The achievement function is essentially an auxiliary function that measures the deviation from the targets set by the DM. The total deviation is not a value of much relevance to the DM as it may be more common (objective) functions as total profit or total cost.

In Goal Programming there are two classical models: the Preemptive model and the Archimedean model. They differ in how they deal with DM's priorities over goals. The Preemptive GP model can be seen as a lexicographic approach as it considers one goal at a time, going from the one with the highest priority to the one with the lowest (Tamiz and Jones 1997). This approach transforms the GP problem into a sequence of single objective linear models. The Archimedean model can be seen as a weighted sum approach, rendering the GP problem as a one single objective linear model.

11.3.1 Preemptive GP Model

In the Preemptive Goal Programming model, one takes the goal with the highest priority for the DM and intersects it with the feasible region. If the resulting set is empty, the compromise solution will be the feasible solution with minimal deviation to the target. Otherwise, an infinite number of solutions are available to be evaluated by the second priority goal. By imposing this second priority goal, one gets a new set of feasible solutions that meet the first and second goals simultaneously. Again, this set may be empty or may have an infinite number of solutions. In the former case, one should compute the feasible solution (amongst the ones that meet the first target) that minimizes the deviation to this second goal. This solution will be the solution the DM is searching for. In the latter case, one should take the third priority goal and proceed as for previous goals. This procedure stops when either the goal has led to an empty set of feasible solutions or all targets have been imposed. The underlying assumption in the Preemptive GP model is that the goal with highest priority to the DM is infinitely more important than the second highest priority goal, and the goal with the second highest priority is infinitely more important than the third highest priority, and so on.

This was the approach followed in Example 11.3. However, no model formulation has been provided. Let's take it again and formulate the problem as it is being solved.

Example 11.4

This GP problem has three goals for $(x, y) \in S$.

- Goal 1: $Z_1 = 20x + 10y$ should not be greater than 650, i.e., $20x + 10y \leq 650$

- Goal 2: $Z_2 = -80x + 40y$ should be smaller or equal to 400, i.e., $-80x + 40y \leq 400$, and
- Goal 3: $Z_3 = y$ should be at between 45 and 65, i.e., $45 \leq y \leq 65$.

The Preemptive GP model formulation for this problem is:

$$\text{Min } G(d_1^+; d_2^+; d_3'^- + d_3'^+)$$
$$\text{s.t. } 20x + 10y + d_1^- - d_1^+ = 650$$
$$-80x + 40y + d_2^- - d_2^+ = 400$$
$$y + d_3'^- - d_3'^+ = 45$$
$$y + d_3^- - d_3^+ = 65$$
$$x + y \geq 20$$
$$2x - y \leq 40$$
$$-x + y \leq 80$$
$$x + y \leq 80$$
$$x, y \geq 0$$
$$x, y, d_1^-, d_1^+, d_2^-, d_2^+, d_3^-, d_3^+, d_3'^-, d_3'^+ \geq 0.$$

Remember that the DM gives the highest priority to goal 1, then to goal 2 and finally to goal 3. So, the first model to solve is

$$\text{Min } d_1^+$$
$$\text{s.t. } 20x + 10y + d_1^- - d_1^+ = 650$$
$$x + y \geq 20$$
$$2x - y \leq 40$$
$$-x + y \leq 80$$
$$x + y \leq 80$$
$$x, y, d_1^-, d_1^+ \geq 0.$$

As it has been seen above, this problem as alternative optima (see Fig. 11.7) and hence the optimal value is zero, $d_1^{+*} = 0$. Note, however, that goal 1 is not equal to 650 but less or equal than it. Therefore, one should allow, in the search space of the second priority goal, solutions that do not exactly meet this first target. In the other order, the constraint modelling goal 1 should be $20x + 10y \leq 650$ and not $20x + 10y = 650$. The model for second priority goal is then formulated as:

$$\text{Min } d_2^+$$
$$\text{s.t. } 20x + 10y \leq 650$$
$$-80x + 40y + d_2^- - d_2^+ = 400$$
$$x + y \geq 20$$
$$2x - y \leq 40$$
$$-x + y \leq 80$$
$$x + y \leq 80$$
$$x, y, d_3^-, d_3^+, d_3'^-, d_3'^+ \geq 0.$$

Again, the optimal value for this second model is zero, $d_2^{+*} = 0$, since there are plenty of solutions that meet both goals (see Fig. 11.8). The model for the third goal can now be formulated as

$$\text{Min } d_3'^- + d_3^+$$
$$\text{s.t.} \qquad 20x + 10y \leq 650$$
$$-80x + 40y \leq 400$$
$$y + d_3'^- - d_3'^+ = 45$$
$$y + d_3^- - d_3^+ = 65$$
$$x + y \geq 20$$
$$2x - y \leq 40$$
$$x + y \leq 80$$
$$x + y \leq 80$$
$$x, y, d_2^-, d_2^+ \geq 0.$$

The optimal solution is $(x, y) = (13.75, 37.5)$ and the optimal deviation from goal 3 is 7.5 as $d_3'^{-*} = 7.5$ and $d_3'^{+*} = 0$. Note that goal 3 is composed of two constraints but only one is not met ($y \geq 45$).

A generic Preemptive GP model can be mathematically formulated as:

$$\text{Min } G_i = P_i(a_i^- d_i^- + a_i^+ d_i^+)$$
$$\scriptstyle i=1,\dots,m$$
$$\text{s.t. } Z_i(x) + d_i^- - d_i^+ = t_i, i = 1, \dots, m$$
$$d_i^-, d_i^+ \geq 0$$
$$x \in S$$

where P_i, $P_i \gg P_{i+1} \gg P_{i+2} \gg \dots$, are the preemptive priority factors that serve only as a ranking symbol (they **ARE NOT** coefficients which multiply the corresponding deviation variables); $a_i^-, a_i^+ \in \{0, 1\}$ are parameters allowing for the presence/absence of the deviation variables in the achievement function.

There are problems where some goals may be classified by the DM as having the same level of priority. In such cases, the "tied" goals should be considered simultaneously (at the same lexicographic level) and the sum of the corresponding undesirable deviation variables minimized. The most common approach is to apply the Archimedean model (Section 11.3.2). However, some care must be taken if goals are of different scales (so that one does not *overshadow* the other(s)). Normalization techniques should then be applied to avoid the *heaviest* deviation variable to dominate the minimization direction.

11.3.2 The Archimedean GP Model

The Archimedean GP model is a weighted sum based model where one seeks to find the feasible solution that is closer to the utopian set. Positive weights are assigned to each goal so that the undesirable deviations are penalized in the achievement function. However, the nature of these weights is quite different from the ones studied in the Weighted Sum model in chapter 9. The GP programming weights aim at minimizing as much as possible the deviation variables of the most important(s) goal(s), allowing for larger deviations in less important goals. These weights do not sum up 1 and, therefore, are not of compensatory nature (as the ones in the Weighted Sum model). However, this modelling approach is especially adequate when the DM wishes to investigate trade-offs among deviation variables (Steuer 1986).

In general, GP models have a large number of deviation variables. If one adds goals with different scale units, assigning weights can be very hard. So, one should start by defining what seems to be a reasonable set of weights and, after analyzing the resulting solution, revise them to see if better[2] solutions can be obtained. Notice, the Archimedean modelling approach renders the GP problem as a single objective linear model and, therefore, can be solved by any software for linear optimization.

Let's take again Example 11.3 and now formulate how to solve it using the Archimedean GP model.

Example 11.5

Remember this GP problem with three goals for $(x, y) \in S$.

- Goal 1: $Z_1 = 20x + 10y$ should not be greater than 650, i.e., $20x + 10y \leq 650$,
- Goal 2: $Z_2 = -80x + 40y$ should be smaller or equal to 400, i.e., $-80x + 40y \leq 400$, and
- Goal 3: $Z_3 = y$ should be at between 45 and 65, i.e., $45 \leq y \leq 65$.

Remember also the DM has the following priorities with respect to the goals:

$$Z_1 \gg Z_2 \gg Z_3.$$

The Archimedean model formulation for this GP problem is

$$\text{Min } G = w_1 d_1^+ + w_2 d_2^+ + w_3 (d_3'^- + d_3^+)$$
$$\text{s.t. } 20x + 10y + d_1^- - d_1^+ = 650$$
$$-80x + 40y + d_2^- - d_2^+ = 400$$
$$y + d_3'^- - d_3'^+ = 45$$
$$y + d_3^- - d_3^+ = 65$$
$$x + y \geq 20$$
$$2x - y \leq 40$$
$$-x + y \leq 80$$
$$x + y \leq 80$$

$$x, y, d_1^-, d_1^+, d_2^-, d_2^+, d_3^-, d_3^+, d_3'^-, d_3'^+ \geq 0.$$

---

[2] "Better" means solutions with the smallest deviations for the highest priority goals.

Since w_i, $i = 1, \ldots ,3$, are the weights modelling the DM's priorities, these should verify

$$w_1 > w_2 > w_3.$$

The vector $(w_1, w_2, w_3) = (3, 2, 1)$ is a set of possible values for the weights. Solving the model with these, one obtains the optimal solution $(x, y) = (13.75, 37.5)$. This solution reflects the DM preferences regarding the goals since $d_1^+ = d_2^+ = d_3^+ = 0$ and only one of the lowest priority deviation variables takes a value different from zero, $d_3'^- = 7.5$.

In the previous example the first set of weight values (all within the same unit range) led to a good solution. However, quite often such weights do not lead to *good* solutions, i.e., solutions that follow the DM's preferences.

Example 11.6

Consider the three goals below defined in set $S = \{(x, y) \in \mathbb{R}^2 : 0 \le x \le 10 \text{ and } 0 \le y \le 10\}$.

- Goal 1: $Z_1 = 2x + 3y$ should be greater or equal to 44,
- Goal 2: $Z_2 = -x + 7y$ should be less or equal to 7, and
- Goal 3: $Z_3 = 5x - y$ should not be greater than 10.

The formulation for the Archimedean GP model is

$$\text{Min } G = w_1 d_1^- + w_2 d_2^+ + w_3 d_3^+$$
$$\text{s.t. } 2x + 3y + d_1^- - d_1^+ = 44$$
$$-x + 7y + d_2^- - d_2^+ = 7$$
$$5x - y + d_3^- - d_3^+ = 10$$
$$0 \le x \le 10$$
$$0 \le y \le 10$$
$$d_1^-, d_1^+, d_2^-, d_2^+, d_3^-, d_3^+ \ge 0.$$

The DM considers goal 1 as the most important and, therefore, the one with the highest priority. Then, (s)he would like to see goal 3 met and goal 2 is the one with the lowest priority for the DM. Hence, the three goals can be ordered as

$$Z_1 \gg Z_3 \gg Z_2.$$

The weight vector $(w_1, w_2, w_3) = (3, 1, 2)$ is in accordance with the priorities given by the DM. With them, the optimal solution is $(x, y) = (10, 4)$ with the following (undesirable) deviations to the targets $d_1^- = 6, d_2^+ = 9$ and $d_3^+ = 0$. These values "say" that this solution is not a *good* one since the goal with the highest priority, goal 1, has 9 units of undesirable deviation while the second most important goal (goal 3) has none. So, a different set of weights should be tested to investigate whether a solution exists that has zero deviation from the goal 1 target. With $(w_1, w_2, w_3) = (30, 1, 2)$ the optimal solution is $(x, y) = (10, 7)$ and the undesirable deviations now $d_1^- = 0$, $d_2^+ = 6$ and $d_3^+ = 15$. This solution is more in accordance with the DM's priorities over the goals. Hence it is a better solution than the first one computed.

A generic Archimedean GP model can be mathematically formulated as:

$$\text{Min } G = \sum_{i=1}^{m} w_i^- d_i^- + w_i^+ d_i^+$$

$$\text{s.t.} \quad Z_i(x) + d_i^- - d_i^+ = t_i, i = 1, \dots, m$$

$$d_i^-, d_i^+ \geq 0, i = 1, \dots, m$$

$$x \in S$$

where $w_i^-, w_i^+ \geq 0, i = 1, \dots, m,$ are the weights assigned to the negative and positive deviation variables, respectively. These weights can be any real number, where the greater the weight the higher is the priority the DM assigns to the respective goal.

As proposed in the example, a sensitivity analysis should be made to weight the values to investigate other solutions that might better please the DM. Jones and Tamiz (2010) proposed an algorithm that performs a complete analysis of the weight space having the DM's initial suggestion as a starting point. Later on, Jones (2011) revised this weight sensitivity algorithm so that it only explores the portion of the weight space of interest to the DM. The revised algorithm may also be used as an interactive tool allowing the DM to explore neighboring solutions in order to reach a solution in line with her/his preferences.

Weights fulfill a second role in this GP model. They may act also as normalization factors, scaling the deviation variables so their values are of the same magnitude. In fact, as pointed out by Jones and Tamiz (2010), the GP Archimedean model weights fulfill a twofold role: they model priorities by establishing the relative importance among goals and act as normalization factors. The authors suggest different normalization strategies: percentage normalization (dividing, in the achievement function each unwanted deviation variable by the target level), zero-one normalization (dividing each unwanted deviation variable by the worst possible deviation within the feasible region; this worst possible value is determined by, independently, solving an LP problem for each goal); and Euclidean normalization (dividing each unwanted deviation variable by the Euclidean norm). All these strategies are applied to the unwanted deviation variables at the achievement function. In Jones and Tamiz (2010), a very educational section is presented on this topic. The interested reader should refer to this work for all the details.

11.3.3 A GP Model Combining Preemptive Priorities and Weighting

The two previous models may be viewed as the two extreme types of GP models. Having them as a baseline, near all the other GP models can be obtained. As

272   Mathematical Models for Decision Making

an example, one can straightforwardly combine the Preemptive model with the Archimedean. According to Schniederjans (1995), this combined model was first proposed by Ijiri (1965). It is particularly suitable when the DM considers several goals to have equal priorities. So, the first step is to ask the DM to group all goals into classes (each having one or more goals) and to order the classes from the highest to the lowest priority. Within each class with more than one goal, the DM is then asked about the weights for the undesirable deviation variables. Note that if all goals are placed in the same class then one has the Archimedean GP model. On the contrary, if there are as many classes as goals, then the model is the Preemptive GP.

A generic combined GP model can be mathematically formulated as:

$$\operatorname*{Min}_{i=1,\dots,m} G_i = P_i \left(\sum_{k=1}^{n_i} w_{ik}^- d_i^- + w_{ik}^+ d_i^+ \right)$$
$$\text{s.t. } Z_i(x) + d_i^- - d_i^+ = t_i, i = 1, \dots, m$$
$$d_i^-, d_i^+ \geq 0, i = 1, \dots, m$$
$$x \in S$$

where $w_{ik}^-, w_{ik}^+ \geq 0, i = 1, \dots, m$, are the positive weights assigned to each of the $k = 1, \dots n_i$ different classes within the ith category, n_i the number of goals assigned to class I, and P_i are the preemptive priority factors.

11.3.4 The Tchebychev GP Model

The Tchebychev or MinMax GP model differs from the Preemptive and the Archimedean models as it makes use of the Tchebychev distance function (L_∞) instead of being based on Manhattan (L_1) distance[3]. Being a MinMax modelling approach, the optimal solution will ensure a balance between the satisfaction of all the goals. Therefore, this approach is the most appropriate one to use when the DM does not seek for goals with zero deviation and others with very high deviations from the targets. This is especially the case in problems where there are multiple DMs, each of whom has a preference to their own subset of goals that they regard as most important (Jones and Tamiz 2010).

A generic Tchebychev GP model can be mathematically formulated as:

$$\text{Min } G = \lambda$$
$$\text{s.t. } Z_i(x) + d_i^- - d_i^+ = t_i, \quad i = 1, \dots, m$$
$$a_i^- d_i^- + a_i^+ d_i^+ \leq \lambda, \quad i = 1, \dots, m$$
$$d_i^-, d_i^+ \geq 0, \quad i = 1, \dots, m$$
$$x \in S$$

---

[3] For more details about distance functions refer to chapter 9.

where $a_i^-, a_i^+ \in \{0, 1\}$ allows for the presence/absence of the deviation variables from the achievement function, and λ is the maximal deviation from amongst the goals.

11.4 Final Remarks

A GP model is a distance-based approach that optimizes multiple goals by minimizing the undesirable deviations from target levels (or goals) set by the DM. An optimal solution where (some of) the undesirable deviations are zero indicates that (some of) the DM's target levels have been achieved (Colapinto et al. 2017).

A very large number of models for GP problem formulations have been proposed over the years. The great majority of these models are a variant or a mix of the baseline models: the Preemptive model and the Archimedean model. In this chapter, these two models are addressed in detail and illustrated by tailored examples. The Preemptive approach models the GP problem through a lexicographic approach. Therefore, it should be used when the DM can easily establish the priority levels or when (s)he does not wish to make direct trade-offs between goals. The Archimedean model is a weighted sum model which allows the study of trade-offs between goals by varying the weights on undesirable deviation variables (Jones and Tamiz 2016). Two additional models are briefly presented exemplifying on how most of the GP models are a mix or a variant of the classical models: combining preemptive priorities and weighting, and the Tchebychev GP model.

Although GP is an apparently easy methodology to apply, it may be misleading when used without a comprehensive knowledge of its basic concepts (Romero 1991). Many of GP shortcomings have been exposed shortly after its appearance. Three very important limitations are:

- Possibility of obtaining dominated solutions as compromise solutions. In Example 11.1, the set of solutions satisfying all goals lies inside the feasible region. In the most common multiobjective problems (as the ones studied in chapters 8 to 10), these solutions were classified as dominated solutions. Only those belonging to the feasible region frontier were candidates to be non-dominated solutions (viewed as *good* solutions). However, remember that to assess whether a solution is dominated or not, one needs to consider all objective functions. In GP there are no objective functions (only the minimization of the undesirable deviations in the achievement function). Romero (1991) states that "GP was not invented with the purpose of obtaining nondominated solutions but was developed as a method for finding satisfactory solutions for complex real world problems" (page 14). Nonetheless, different procedures have been proposed to overcome this issue (e.g. Romero (1991) and Jones and Tamiz (2010)).

- In the Preemptive GP model, the use of excessive numbers of priority levels may lead to the redundancy of the lower priority goals. In other words, only when there are alternative optima, the optimization process may proceed to the next priority level. Therefore, if no alternative optimal solution exists for the problem which corresponds to the ith priority level, then the goals placed in priorities lower than i^{th} will not be able to produce a different result. Consequently, they are redundant. Amador and Romero (1989) addressed this issue extensively.

- Direct comparison of incommensurable goals. In the Archimedean GP model, being a weighted sum model, one might mix "apples with pears". For instance, suppose one is a budgetary goal measured in monetary terms, while the other goal may be a human resource goal measured in terms of workers assigned to a task. Both goals are expressed in different and unrelated units. One strategy to deal with this issue is the use of scaling methods that will normalize the goals (Romero and Rehman 1984). Another strategy is to avoid the Archimedean modelling approach and to apply the Preemptive model if the DM agrees to prioritize the goals. In this latter model, incommensurability is not an issue.

As mentioned in the introductory section, this chapter aims at being a first step into the very large research area that is GP modelling. The interested reader, wishing to deepen his/her knowledge about GP, should refer to dedicated books on the topic as Jones and Tamiz (2010), Schniederjans (1995), or Romero (1991). Also, the limitations of GP modelling have been extensively addressed. Jones and Tamiz (2010) and Romero (1991) are two very comprehensive works on these matters.

11.5 Proposed Exercises

1. Consider the GP problem
   $G_1 : x$ should be greater or equal to 8,
   $G_2 : y$ should be greater or equal to 9.
   $$\text{s.t.} \quad 3x + y \leq 24$$
   $$2x + 7y \leq 35$$
   $$x, y \geq 0$$

   (a) Graph set S and the utopian set.
   (b) Assuming both goals have equal priority, determine the optimal solution that minimizes the sum of the undesirable deviations.
   (c) Determine the solution in S that optimizes the Preemptive model where goal 1 is of higher priority than goal 2. Will the optimal solution change if goals are ordered differently?

2. Let S be the feasible region defined by the constraints below.

$$-x + y \leq 3$$
$$x + y \leq 8$$
$$x \leq 6$$
$$y \leq 4$$
$$x, y \geq 0.$$

Consider the goals below with the priority order: $G_1 \gg G_2 \gg G_3$.
$G_1 : x$ should not exceed 3 units,
$G_2 : 3x - y$ should be greater or equal to 12 units,
$G_3 : y/2$ should be equal to 1 unit.

(a) Propose a formulation for the Preemptive Goal Programming model. Determine the optimal solution.
(b) Propose a formulation for the Archimedean Goal Programming model. Use an optimization software to compute the optimal solution.

3. Let S be the feasible region defined by the constraints below.

$$-x + 2y \leq 24$$
$$x + 4y \leq 84$$
$$y \geq 4$$
$$x \leq 20$$
$$5x + 2y \geq 48$$
$$x, y \geq 0.$$

Consider the following goals:
$G_1 : x + y$ should not be less than 20 nor greater than 25
$G_2 : 4x - 2y$ should not exceed -5, and
$G_3 : x$ should not be less than 15.

(a) Using the Archimedean GP propose a model that formulates the problem of a DM that gives the highest priority to goal 3, followed by goal 1.
(b) Using the priorities of the previous exercise, apply the Preemptive GP model to determine the optimal solution. Compute the deviation concerning each goal.
(c) Assume the DM has changed her priorities with respect to the goals. Justify the statement: "An optimal solution previously computed will not hold as the optimal solution".

4. Consider the GP problem

$G_1 : 3x + y$ should be between 40 and 48 units

$G_2 : -2x + y$ should not be negative

$G_3 : -x - 4y$ should be less than -45 units

$$\text{s.t.} \quad x + 2y \leq 27$$
$$-2x + 5y \geq 0$$
$$x + y \leq 14$$
$$x, y \geq 0$$

(a) Represent graphically the feasible region and the utopian set (if exist).

(b) Propose a formulation for the Preemptive GP model considering $G_2 \gg G_3 \gg G_1$. Do altering goal priorities have an impact on the optimal solution? Justify the answer.

(c) Propose a formulation for the Archimedean GP model that reflects the following DM: "I do not wish to have large deviations from target 2, but I do not mind if target 1 is not met."

5. Consider GP problem goals below with the priority order: $G_1 \gg G_2 \gg G_3$.

$G_1 : x + y$ should not less than 4 nor higher than 6,

$G_2 : 4x - y$ should not exceed 6,

$G_3 : x$ should be greater or equal to 5.

$$\text{s.t.} \quad x + 3y \geq 9$$
$$-2x + y \leq 4$$
$$2x + y \leq 14$$
$$x, y \geq 0.$$

(a) Depict graphically the feasible region and the utopian set (if exists).

(b) Knowing the DM prioritize goal 3 over goal 1 and goal 1 over goal 2,

(i) propose a formulation for the Archimedean GP model,

(ii) solve the problem graphically using the Preemptive modelling approach.

6. Two types of ink are produced in a factory, A and B. Each kiloliter of ink generates a profit of 1 and 2 monetary units for A and B, respectively. It is only possible to produce a total of 120 kiloliters of paint a day. Unfortunately, ink production creates toxic waste. Waste quantities are 3 kg of Alpha waste per kiloliter of ink A produced and 5 kg of Beta waste per kilogram of ink B. There are two recycling systems for treating Alpha and Beta waste, each with a daily capacity to process 225 kg of waste. Each load of waste must be handled separately and whenever the capacities of the recycling systems are exceeded it is necessary to pay a company to deal with the surplus.

It is desirable that:

- the total daily profit is greater than 180 monetary units,
- the capacity of each of the two recycling systems is not exceeded,
- and the production of A is half of the production of B.

(a) What is the meaning of each of the expressions below?

$$x + y \geq 180$$
$$-x + y \leq 120$$
$$x \leq 75$$
$$2x - y = 0$$
$$y \leq 45$$

(b) Formulate the problem as a goal programming model.

(c) Solve the problem graphically and according to a following priority ordering:
  (i) 1, 2 e 3
  (ii) 2, 3 e 1.

(d) Applying the Archimedean method and considering the goals order 1, 2 e 3, determine the solution with the weights:
  (i) 3, 2 e 1
  (ii) 300, 20, 1

(e) Compare and comment on the two solutions obtained in the previous question.

7. Aunt Leonarda's assistants, Bernardo and Zumélia, enthusiastically discuss the plan for the next week's production of their small perfume factory. They have several problems to solve, the biggest of one is the absence of the boss.

"What a bummer! Why did Aunt have to go to Madagascar right now??" lamented Zumélia. "I know that the rarest essences are found in these hidden places, but I have no idea how to manage this business."

"Everything is going beautifully, my dear, "Bernardo tried to calm her down." You're doing great. I already understood this memo left by Aunt Leonarda. To produce a kiloliter of *Ambarine*, our new perfume, we need 4 kg of Xylomenadrine and 2 kg of Ymutiridamine, while for one kiloliter of the classic *Bliss* we only need 2 kg of each ingredient."

"Well, you are being very optimistic. Don't you see that there is only 60 kilos of 'Xylo... something' and 40 kilos of 'Ymuriri' in the warehouse!? I will order more, but the order will not arrive in time for this week production." Zumélia grumbled.

"Do not forget that Aunt Leonarda wishes to produce at least 6 kiloliters of each of the perfumes ... When any of them is below that value she turns into a beast! If I know her well, I estimate that the number of outbursts that we are going to be putting up with will be proportional to the number of kiloliters we produce below 6. Do you remember what happened two Christmas ago?", whispered Bernardo.

"Oh! I do!! Last month it happened once again, but it was not so terrible." replied Zu, remembering the last Aunt's fury. But I still think that when failures occur with *Bliss*, Aunt's level of exasperation is twice as high as when failures occur with *Ambarine*.

"Let me change the subject, Aunt Leonarda left me a note, asking to be careful with the production costs. I cannot read the number, but it seems to me that it is preferable that the costs are kept lower than 36000 N (neuros)."

"From my calculations, I estimate that producing a kiloliter of *Ambarine* and *Bliss* costs, respectively, 6000 and 4000 N, noted Zumélia , "But I'm almost certain that Auntie usually prefers costs lower than 72 thousand N."

"Oh, it really looks like a 36!" replied Bernardo, narrowing his eyes.

"And what do you think is more important? To produce more than 6 kl of *Bliss* and more than 6 kl of *Ambarine*, or to be careful with the costs?", asked Zumélia patiently.

"I have no idea!! Is it relevant?", asked Bernardo.

(a) Apply the Preemptive Goal Programming Model,
  (i) Assuming that Aunt Leonarda's note indicates 72 000 N, investigate whether it is relevant to know the most important objective.
  (ii) Will the conclusions drawn on the previous question still hold if, after all, the costs should be below 36 000 N?
(b) Solve exercise (a.ii) using the Archimedean model.

8. Bernardo is planning his study for the next week and decided to do it "scientifically". He needs to decide how many hours to study Topology and Operational Research. Bernardo would very much like his total study time next week not to exceed 24 hours, after all he needs to sleep, eat calmly and spend some time with his friends. Ideally, the number of study hours for Topology would be approximately half the number of hours of OR. Otherwise, he is not able to withstand so much Topological Space. And, finally, it would be very good to study at least 10 hours of Topology, otherwise Bernardo is afraid of a bad grade. Help Bernardo to plan his study under the following conditions:

(a) If the three goals are of equal importance. However, in this scenario, deviations from the second goal are doubly more annoying than deviations from the remaining two goals.
(b) When the priority of the goals is the same as the order in which they were described.
(c) If the priority order is reversed concerning how they have been presented.

9. Consider region S analytically defined by the equation system:

$$x + y + d_1^- - d_1^+ = 14$$
$$y + d_2^- - d_2^+ = 10$$
$$y + d_3^- - d_3^+ = 45$$
$$x - y + d_4^- - d_4^+ = 7$$
$$x, y, d_1^-, d_1^+, d_2^-, d_2^+, d_3^-, d_3^+, d_4^-, d_4^+ \geq 0.$$

Solve the following GP problems:

(a) *Min* $G = (d_1^- + d_1^+; d_3^-)$ s.t. S
(b) *Min* $G = (d_2^-; d_3^- + 2d_4^-)$ s.t. S
(c) *Min* $G = (d_1^-; d_4^+; d_2^+; d_3^-)$ s.t. S

10. Consider the region R represented by the following system of equations:

$$x + y + d_1^- - d_1^+ = 14$$
$$y + d_2^- - d_2^+ = 10$$
$$y + d_3^- - d_3^+ = 45$$
$$x - y + d_4^- - d_4^+ = \alpha$$
$$d_1^-, d_1^+, d_2^-, d_2^+, d_3^-, d_3^+, d_4^-, d_4^+ \geq 0.$$

with $\alpha \in R$.

(a) Assume $\alpha = 2$. Indicate whether the priorities assigned to each goal are relevant to achieve an optimal solution to any Goal Programming Problem in region R. Justify your answer.

(b) Assume $\alpha = 5$. Indicate, if possible, the points that are the solution following the Goal Programming problems. Note, there may be more than one solution (tick all the possible solutions) or none of the points is a solution (in this case tick "none").

(i) *Min* $(d_3^- + d_2^+; d_1^+; d_4^-)$

☐ (4, 10) ☐ (8, 6) ☐ (11, 6) ☐ (15, 10) ☐ None

(ii) *Min* $(d_3^-; d_1^-; d_4^-+ d_4^+)$

☐ (4, 10) ☐ (8, 6) ☐ (11, 6) ☐ (15, 10) ☐ None

(iii) *Min* $(d_4^+; d_3^- + 4d_2^-)$

☐ (4, 10) ☐ (8, 6) ☐ (11, 6) ☐ (15, 10) ☐ None

11. Apply the Preemptive GP model to solve the problem:

G_1 : x should be greater or equal to 5,

G_2 : y should be greater or equal to 3,

G_3 : z should be greater or equal to 4,

$$\text{s.t.} \quad 2x + 3y + 2z \leq 16$$
$$x + 2y \leq 5$$
$$x, y, z \geq 0$$

Assuming:

(a) $G_1 \gg G_2 \gg G_3$

(b) $G_2 \gg G_1 \gg G_3$

(c) Provide some comment concerning the impact priories have on the optimal solution.

12. One of the most important problems in statistics is linear regression. Briefly, this problem can be defined as determining the coefficients of a straight line that better fit a set of data represented by points, (x_1, y_1), (x_2, y_2), ..., (x_n, y_n). In other words, if the line is represented by $y = a + bx$, the objective will be to determine constants $a, b \in R$ that minimize the deviation according to a certain criterion. This criterion is usually called the least squares method. However, other interesting criteria can be use so that a linear programming model can be applied to determine the values of constants a and b. For example, to minimize the sum of absolute deviations, i.e.,

$$Min \sum_{i=1}^{n} |y_i - (a + bx_i)|.$$

Formulate the linear regression problem with the criterion of the sum of the absolute deviations as an Archimedean Goal Programming model.

References

Amador, F. and C. Romero (1989). Redundancy in lexicographic goal programming: An empirical approach. *European Journal of Operational Research*, 41(3), 347–354.

Charnes, A., W.W. Cooper and R. Ferguson (1955). Optimal estimation of executive compensation by linear programming. *Management Science*, 1, 138–151.

Charnes, A. and W.W. Cooper (1961) *Management Models and Industrial Applications of Linear Programming*. Wiley, New York.

Colapinto, C., R. Jayaraman and S. Marsiglio (2017). Multi-criteria decision analysis with goal programming in engineering, management and social sciences: A state-of-the art review. *Annals of Operations Research*, 251(1-2), 7–40.

Ijiri, Y. (1965). *Management Goals and Accounting for Control*. North Holland Pub. Co.

Jones, D. (2011). A practical weight sensitivity algorithm for goal and multiple objective programming. *European Journal of Operational Research*, 213(1), 238–245.

Jones, D. and M. Tamiz (2010). *Practical Goal Programming* (Vol. 141). New York: Springer.

Jones, D. and M. Tamiz (2016). A review of goal programming. *In:* Multiple Criteria Decision Analysis (pp. 903–926). Springer, New York, NY.

Romero, C. (1991). *Handbook of Critical Issues in Goal Programming.* Pergamon Press, Oxford.

Romero, C. and T. Rehman (1984). Goal programming and multiple criteria decision-making in farm planning: An expository analysis. *Journal of Agricultural Economics,* 35(2), 177–190.

Schniederjans, M. (1995). *Goal Programming: Methodology and Applications.* Springer Science & Business Media.

Steuer, R.E. (1986). *Multiple Criteria Optimization: Theory, Computation and Applications.* John Wiley and Sons, Inc.

Tamiz, M. and D.F. Jones (1997). Interactive frameworks for investigation of goal programming models: Theory and practice. *Journal of Multi-Criteria Decision Analysis,* 6(1), 52–60.

Wierzbicki, A.P. (1982). A mathematical basis for satisfying decision making. *Mathematical Modelling,* 3(5), 391–405.

Appendix B: Definitions

A.1 Multi-Criteria Methods

Chapters 3 to 6 will present different kinds of Multi-Criteria Decision methodologies but some of these important concepts are common to all of them. They will be presented here so that the reader can easily follow the four chapters concerning MCD techniques. A simple example will also be presented. Although it will be thoroughly explored in each MCD chapter, the example will also appear here to frame the presented concepts and definitions.

A **Decision Maker** (DM) is the entity (or entities) responsible for the final word on a Decision Making process. It is usually a person, or a small number of persons, but can also be an institution. Even then, there is always someone that represents the institution's interests and will be responsible for all judgments' during the decision making process.

An **Alternative** is one of the possible choices presented to the DM. In a Multi-Criteria problem the set of possible Alternatives is always finite and easily identifiable. The Alternatives have to be identified and discriminated against by the DM.

An **Attribute** is a measure that allows the DM to evaluate each available Alternative according to a particular point of view. It can be a numerical Attribute (e.g. length, weight, a grade of some scale, or distance to a specific place) or a qualitative Attribute (e.g. comfort, pleasantness, location, amongst many others). These latter Attributes are usually very hard to quantify. Consequently, they are usually measured by a qualitative scale (e.g. "Excellent", "Very Good", "Good") representing the DM's subjective biases. In some methods Attributes and Criteria are equivalently used, although sometimes the terms are correlated but they do not represent exactly the same concept.

A.1.1 An Example

The following example will be used as a framework for all four Multi-Criteria Methodologies presented in Chapters 4 to 6.

The Team Leader (TL) of a research group was assigned with the responsibility of hiring a new technician for the team. The Science Lab where the group does its research work, wants to hire someone likeable, with good communication skills, efficient, effective and knowledgeable. Several candidates applied for the place and Human Resources has already selected who they considered to be the most qualified. The TL thinks that it will be easier to assess the candidates' likeability and communications skills by interviewing the candidates. These interviews will be rated as "Very Weak", "Weak", "Regular", "Good", "Very Good" and "Excellent". The efficiency and effectiveness will be measured through a practical test with several common lab tasks that have to be correctly executed in a small amount of time. Each non executed or flawed task will penalize the candidate with a certain number of demerits. More than 10 demerits in the practical test will imply the elimination of a candidate. Finally, the candidates' knowledge will be assessed through a theoretical admission exam, rated from 0 to 100%.

Table A.1 presents the results of the selected candidates who passed the practical test.

Table A.1: Example's Data

		Interview	Demerits in Practical Test	Theoretical Test
	A	Very Good	4	75.4 %
Candidates	B	Good	2	84.6 %
	C	Regular	3	95.7 %
	D	Good	6	90.2 %

In the presented example the Decision Maker will be the Team Leader that will be responsible for picking the most adequate candidate. The candidates will be the Alternatives. The Attributes/Criteria will be the Interview evaluation, the number of demerits obtained by a candidate during the practical test and the grade of the Theoretical Test.

A.2 Multi-Objective Methods

As for multi-criteria methods, multi-objective methods share several concepts. The definitions given below are presented in detail in chapter 7 and illustrated with some examples. However, for an easier reference when reading chapters 8 to 11, they are presented in this chapter in their short version.

Given a feasible space S, a point $x^* \in S$ is called an **efficient solution** if and only if there is no other solution $x \in S$ such that

(i) $Z_i(x) \geq Z_i(x^*)$, $i = 1, ..., k$

(ii) $\exists_{i = 1, ..., k} Z_i(x) > Z_i(x^*)$.

The set of all efficient solutions is known as **Efficient Frontier** and is represented by S^e.

Given a feasible space S, a point $x^* \in S$ is called **weakly efficient solution** if and only if there is no other solution $x \in S$ such that

$$Z_i(x) \geq Z_i(x^*), \; i = 1, ..., k$$

The objective functions space or, more simply, as **objectives space** is the space defined by the images of each feasible solution by each objective function and is represented as S_z. In other words, for each solution $x \in S$ there is a point $z = (Z_1(x), Z_2(x), ..., Z_k(x))$ which is the projection of x in the objective functions space.

A point in the objectives space $z \in S_z$ is called a **non-dominated solution** (or **Pareto optimal solution**) if and only $x \in S$ is an efficient solution with $z = (Z_1(x), Z_2(x), ..., Z_k(x))$. In other words, the set of all non-dominated solutions can be defined as

$$S_z^e = \{z = (Z_1(x), Z_2(x), ..., Z_k(x)) : x \in S^e\}.$$

Set S_z^e is commonly known as **Pareto Front** or **Pareto Frontier**.

A point $z = (Z_1(x), Z_2(x), ..., Z_k(x)) \in S_z$ is a **weakly non-dominated solution** if and only if $x \in S$ is weakly efficient.

The **payoff table** (Table A.2) provides the objective function values with respect to the optimal solution of each objective function when optimized individually. Its first column presents all optimal solutions and the remaining columns the image of these solutions by each objective function. Its generic element z_p^i is the value of the objective function Z_i optimal solution $\left(x_p^i\right)$ by objective function Z_p. Each row corresponds to a solution in the objective functions space. The values on the main diagonal $(Z_p^*, p = 1, ..., k)$ are the optimal value of each objective function.

Table A.2: Payoff Table

	Z_1	Z_2	...	Z_k
x_1^*	Z_1^*	Z_1^2	...	Z_1^k
x_2^*	Z_2^1	Z_2^*	...	Z_2^k
...	...	...	...	...
x_k^*	Z_k^1	Z_k^2	...	Z_k^*

The point in the objectives space $Z^* = (Z_1^*, Z_2^*, ..., Z_k^*)$ is known as the **ideal solution**.

The **nadir solution** is the point whose coordinates are the worst values (the minimum values if all objective functions have maximization as the optimization direction) for the objective functions over the non-dominated solutions:

$$N^* = (n_1^*, n_2^*, ..., n_k^*)$$

where $n_p^* = \min\{Z_j^p : j = 1, ..., k\}, \, p = 1, ..., k$.

Index

For Product Safety Concerns and Information please contact our EU
representative GPSR@taylorandfrancis.com
Taylor & Francis Verlag GmbH, Kaufingerstraße 24, 80331 München, Germany

www.ingramcontent.com/pod-product-compliance
Lightning Source LLC
Chambersburg PA
CBHW060339220326
41598CB00023B/2757

9 781032 168456